Debating Nature's Value

Victor Anderson
Editor

Debating Nature's Value

The Concept of 'Natural Capital'

palgrave
macmillan

Editor
Victor Anderson
Global Sustainability Institute
Anglia Ruskin University
Cambridge, UK

ISBN 978-3-319-99243-3 ISBN 978-3-319-99244-0 (eBook)
https://doi.org/10.1007/978-3-319-99244-0

Library of Congress Control Number: 2018968273

This Palgrave Pivot imprint is published by the registered company Springer Nature Switzerland AG
The registered company address is: Gewerbestrasse 11, 6330 Cham, Switzerland

Acknowledgements

The chapters in this book were written as a result of the work of the Debating Nature's Value Network, funded by the Arts & Humanities Research Council (AHRC; grant number AH/N006232/1). I would like to thank AHRC and my colleagues on the coordinating group for the Network—Felicity Clarke, Rupert Read, and Aled Jones—and everyone who took part in the Network's events.

Contents

Notes on Contributors

Victor Anderson is a Visiting Professor at the Global Sustainability Institute, Anglia Ruskin University. He has previously worked as an economist for the UK Sustainable Development Commission and WWF-UK (World Wildlife Fund), has been an elected member of the London Assembly and was appointed to the Board of the London Development Agency. He is the author of two books, *Energy Efficiency Policies* (Routledge, 1993) and *Alternative Economic Indicators* (Routledge, 1991, reprinted 2013).

Lenke Balint is Head of Communities and Capacity Development at BirdLife International. Working with BirdLife's global network of over 120 national NGO partners, she leads, fundraises for, and promotes two global programmes dealing with strengthening the BirdLife Partnership at national and local level. BirdLife provides crucial support and the fundamental institutional basis for grassroots organisations to effectively conserve, manage, and defend their natural capital in the long term, as well as to ensure the voices of local groups are heard by governments and decision-makers. Balint is also an Intergovernmental Science-Policy Platform on Biodiversity and Ecosystem Services (IPBES) Global Fellow and is an author on IPBES's first Global Assessment on Biodiversity and Ecosystem Services. Through her work with IPBES, she was involved in assessing the status and trends of global biodiversity and ecosystem services, and understanding the impact of biodiversity and ecosystem services on human well-being and the effectiveness of management responses. Balint has a BSc in Agricultural Economics and an MPhil in Conservation Leadership, and is interested in natural capital and its links to sustainable and equitable nature conservation.

Sandra Bell is Lead Nature Campaigner for Friends of the Earth, currently focusing on a campaign for pesticide reduction and a new farming policy that supports nature-friendly farming. This follows having a key role in the organisation's Bee Cause campaign, which successfully worked with a range of allies to convince the UK Government to publish a National Pollinator Strategy and back a tougher ban on bee-harming neonicotinoids. Bell has also worked on an EU-wide campaign to defend European nature laws and on a range of food and farming issues from supermarket power to GM food. Prior to working at Friends of the Earth, Bell had a background in local government environment and planning policy.

Rebecca Clark has worked as an environmental economist for Natural England (the government's advisor on the natural environment in England) for the past ten years. Her work involves applications of environmental economics in marine, freshwater and terrestrial environments, working with researchers, non-government organisations and public sector organisations. She has experience in employing a natural capital approach and recently undertook the study 'Is Corporate Natural Capital Accounting appropriate for monitoring nature reserves?' Previously, Clark worked as a research and teaching fellow in academia.

John Foster is a freelance writer and philosophy teacher and an associate lecturer at Lancaster University. He is the author of *The Sustainability Mirage* (Earthscan, 2008) and *After Sustainability: Denial, Hope, Retrieval* (Routledge, 2015). In addition, he has edited *Valuing Nature? Economics, Ethics and Environment* (Routledge, 1997) and *Post-Sustainability: Tragedy and Transformation* (Routledge, 2018).

Tom Greaves is Senior Lecturer in Philosophy at the University of East Anglia. He works in the area of environmental aesthetics and ecological phenomenology. His recent published work is in the area of animal aesthetics and the concept of nature in phenomenology.

Aled Jones FHEA, HonFIA, is the inaugural Director of the Global Sustainability Institute (GSI) at Anglia Ruskin University. Over the past seven years the GSI has grown into an internationally recognised brand, with a group of 40 individuals, and over £6 million in external income won. He is one of the acknowledged global leaders in public–private finance related to the green economy. His work in climate finance has been recognised by the State of California and he has received a key to the city of North Little Rock, USA.

Dario Kenner is a Visiting Fellow at the Global Sustainability Institute based at Anglia Ruskin University. His chapter is based on the Institute of Chartered Accountants of England and Wales (ICAEW) report "Who should value nature?" (2014). His research focuses on the links between inequality and environmental issues.

Tom H. Oliver is Professor of Applied Ecology at the University of Reading, UK. He is Research Division Leader for their Ecology and Evolution Division and sits on the European Environment Agency scientific committee. His research focuses on understanding the interacting impacts of drivers upon biodiversity and consequent impacts for ecosystem functions and services. A key aspect of this involves developing methods and tools to better quantify and communicate environmental risk to support environmental decision-making.

Jenneth Parker PhD (Sussex), Msc (LSE), BA (Cardiff), Cert Ed, is a research director at The Schumacher Institute for Sustainable Systems based in Bristol. For 10 years she was Co-Director of an international Master's Programme in Education for Sustainability at London South Bank University with students across the world. In addition to her academic qualifications she is a qualified adult and community educator with wide experience in teaching adults and facilitation of participatory events. She has undertaken evaluation, policy and development work for WWF-UK, Local Authorities, UNESCO, the EU and the Welsh Government. Jenneth has facilitated many interdisciplinary research workshops, including at the University of Bristol, managing the synthesis phase of the major Natural Environment Research Council *QUEST* project, designed to add biotic feedbacks into climate change models. She is involved in the EU-funded Marie Curie *Adapt Econ II* project, with 12 PhD students using systems dynamics to map economic change in different economic sectors in the light of planetary and resource limits. She has published widely on sustainability, learning and philosophy, including the book *Critiquing Sustainability, Changing Philosophy*.

Rupert Read is a reader in Philosophy at the University of East Anglia. A specialist in Wittgenstein, he has written and edited a number of influential books on the subject. Aside from this, Read's key research interests are in environmental philosophy and philosophy of film. His research in environmental ethics and economics has included publications on problems of 'natural capital' valuations of nature, as well as pioneering work on the

Precautionary Principle. Recently, his work was cited by the Supreme Court of the Philippines in their landmark decision to ban the cultivation of GM aubergine. Read is also chair of the UK-based post-growth think tank Green House, and is a former Green Party of England and Wales councillor, spokesperson, European parliamentary candidate and national parliamentary candidate. He stood as the Green Party MP-candidate for Cambridge in 2015.

Rob Tinch is an environmental economist working on the economics of biodiversity and ecosystems and associated policy advice. He works primarily on the socio-economic components of European research projects, currently including IMPRESSIONS (modelling adaptation to high-end climate scenarios), ATLAS (spatial planning for the deep Atlantic), and MERCES (coastal ecosystem restoration). He also works on shorter projects for a range of UK and international clients, mostly in the public sector.

Samir Whitaker currently manages the Business and Biodiversity Programme at the Zoological Society of London (ZSL). Working closely with ZSL's Corporate Partnerships team, he is involved in developing strategic partnerships with the corporate sector, mobilising corporate partners to help ZSL achieve its conservation objectives. Prior to joining ZSL, Whitaker first worked on biodiversity management for the private sector, and subsequently with an NGO, BirdLife International, on developing and managing corporate partnerships. Whitaker has an MSc in Environmental Chemistry and an MPhil in Conservation Leadership, and is interested in the assessment and management of natural capital in a variety of sectors and supply chains.

List of Tables

CHAPTER 1

Introduction

Victor Anderson

Abstract This introduction outlines both the book and the work of the Debating Nature's Value Network which the book is based on. It points out the controversial nature of the "natural capital" idea, and briefly outlines how the different chapters approach the issues raised by the use and implications of this concept.

The concepts we use to describe the world are not neutral. They have their own implications, histories, and often supporters and critics. The concept of "natural capital" is one such contested concept, playing a key role in current debates about the future of the natural world. For some, it is seen as a means of protecting the planet, whilst others see it as a threat to lives and livelihoods.

There is, however, one area where it is not much contested. Within the economics community in the UK, it is often simply accepted. It has also

V. Anderson (✉)
Global Sustainability Institute, Anglia Ruskin University, Cambridge, UK
e-mail: Victor.anderson@anglia.ac.uk

V. Anderson (ed.), *Debating Nature's Value*,
https://doi.org/10.1007/978-3-319-99244-0_1

received the support of four major UK ecological science organisations.[1] The term "natural capital" is mentioned 90 times in the Government's latest statement of environmental policy, its 25-Year Environment Plan.[2] This bandwagon has well and truly rolled amongst economists and policy-makers in this country.

Its UK critics are largely to be found amongst the general public, and within academia amongst practitioners of the arts and humanities. It is the Arts and Humanities Research Council (AHRC) that has supported the work that led to this book, through their funding for a network to meet and discuss the issues raised by "natural capital", called Debating Nature's Value (DNV).

The DNV network has held two major meetings: one at the University of East Anglia, in Norwich, in March 2017; the other at Anglia Ruskin University, in Cambridge, in April 2018. One result was the commissioning of a series of papers, some of which are now included in updated versions in this book. A piece of select committee evidence was produced.[3] There is also a website, including an online glossary of terms[4]; and there has been an art exhibition and commissioned artwork created by Rosanna Greaves. There are now hopes of following through our discussions with more detailed work with groups of specialists and decision-makers.

From the start, as organisers we wanted to avoid a simple polarised debate. We wanted to explore nuances and middle positions as well as those at each end of the spectrum. This is reflected in contributions to this book.

We also wanted to reflect the international context of this debate. Whilst "natural capital" has very strong support in the UK, it is far less popular in some other parts of the world. This has resulted in the main

[1] The British Ecological Society, Royal Society of Biology, Centre for Ecology & Hydrology, and the James Hutton Institute formed a partnership centred on the "natural capital" concept, the Natural Capital Initiative.

[2] UK Government: 'A Green Future: Our 25-Year Plan to Improve the Environment' (2018).

[3] Evidence to the House of Commons Environmental Audit Committee inquiry into the Government's 25-Year Plan for the environment: http://data.parliament.uk/writtenevidence/committeeevidence.svc/evidencedocument/environmental-audit-committee/25-year-environment-plan/written/79050.html

[4] DNV website: https://www.anglia.ac.uk/global-sustainability-institute-gsi/research/global-risk-and-resilience/debating-natures-value

international forum for scientific advice to governments about biodiversity—Intergovernmental Science-Policy Platform on Biodiversity and Ecosystem Services (IPBES)—choosing instead to use the far less specific and more inclusive term "Nature's contributions to people".[5] As a global organisation, it would have been in difficulties if it had based its assessments on a concept regarded with suspicion in large parts of the world, especially in Latin America. One aim of this book is therefore to draw attention to this international debate, which has a very different flavour to the debate in the UK.

The chapters in this book start from an account of the international debate about biodiversity and natural capital. Lenke Balint and Aled Jones look particularly at developments in IPBES. This is followed by a critique by Rupert Read and Tom Greaves of the single most globally influential contribution to the promotion of "natural capital" as a concept: the studies carried out by The Economics of Ecosystems and Biodiversity (TEEB). After that, there is an overview of the Natural Capital Protocol by Samir Whitaker, which has taken the sorts of ideas discussed by TEEB out amongst decision-makers, particularly in business.

The next two chapters, by Rob Tinch and Tom Oliver, both set out, after examining the strengths and weaknesses of "natural capital" as an idea, nuanced positions based on taking account of both sides of the argument. Sandra Bell and Rebecca Clark then both consider the practical implications of "natural capital" in the work of Friends of the Earth and English Nature.

Practical issues of a different sort are discussed in the chapters by Victor Anderson and Dario Kenner. I discuss a series of difficulties in arriving at figures for the monetary value of natural capital, whilst Dario discusses the particular difficulty caused by different groups of people valuing nature differently.

The next two chapters, by Jenneth Parker and John Foster, discuss some of the wider issues raised in the natural capital debate, exploring ways of thinking about the natural world which challenge the whole basic approach taken by "natural capital" advocates.

[5] See Unai Pascual et al.: 'Valuing nature's contributions to people: the IPBES approach', in 'Current Opinion in Environmental Sustainability' June 2017, Vols. 26/27, pages 7 – 16. https://www.sciencedirect.com/science/article/pii/S1877343517300040

In the final chapter, however, I end by suggesting that the whole debate about "natural capital" is itself a problem, diverting attention from the real drivers of biodiversity loss and ecosystem deterioration, which above all are the economics of land use change and the economics of climate change. This provides a different approach to what "the economics of biodiversity" might mean: not the application of a mainstream economics concept (capital) to the natural world, but an investigation of economic forces as causes of change. This chapter also serves as a reminder of the global context not only in terms of the debate about concepts, but also in terms of the devastation now happening in the biosphere.

The co-ordinating group for the DNV network consisted of Dr. Rupert Read as Principal Investigator, Dr. Aled Jones as Co-Investigator, Felicity Clarke as Administrator, and myself as Researcher. We are grateful to the many people who gave us advice as the network progressed and to the AHRC for their support for our work.

CHAPTER 2

Natural Capital and the Intergovernmental Science-Policy Platform on Biodiversity and Ecosystem Services

Lenke Balint and Aled Jones

Abstract Recent recognition of the reliance of human economies on natural resources has been a crucial achievement for policymaking. However, there remains a gap in the knowledge of the full extent of the connection between human economies and natural resources. This is relevant for policymaking as understanding who affects the generation of ecosystem services (called 'providers' or 'suppliers') and who benefits from ecosystem services ('beneficiaries' or 'consumers') allows assessments of the costs and benefits from any given policy, including the distributional consequences across affected parties. In this chapter, we explore progress towards furthering this particular gap in knowledge, reflecting on a number of conceptual ecosystem service assessment frameworks developed in the last decade, including the one deployed by the Intergovernmental Science-Policy Platform on Biodiversity and Ecosystem Services, a recently established intergovernmental body; and its efforts to inform policy formulation.

L. Balint (✉)
BirdLife International, Cambridge, UK
e-mail: Lenke.balint@birdlife.org

A. Jones
Global Sustainability Institute, Anglia Ruskin University, Cambridge, UK
e-mail: Aled.Jones@anglia.ac.uk

V. Anderson (ed.), *Debating Nature's Value*,
https://doi.org/10.1007/978-3-319-99244-0_2

Summary

The ongoing degradation and loss of biodiversity and ecosystems have become an inescapable part of reality. Much of it is due to the ways in which humans make use of nature and in the effects of our dominant market-based economic model, which so far has been almost blind to the coupling between the economy and the environment [1]. Recent recognition of the reliance of human economies on natural resources has been a crucial achievement for policymaking. However, there remains a gap in the knowledge of the full extent of the connection between human economies and natural resources.

This is relevant for policymaking as understanding who affects the generation of ecosystem services (called 'providers' or 'suppliers') and who benefits from ecosystem services ('beneficiaries' or 'consumers') allows assessments of the costs and benefits from any given policy, including the distributional consequences across affected parties [2]. In addition, concerns about how ecosystems are able to respond to climate change and other gradual or abrupt changes have led to greater efforts to understand their resilience from local to planetary scales [3] and how interventions can increase or reduce this resilience.

In this chapter, we explore progress towards furthering this particular gap in knowledge, reflecting on a number of conceptual ecosystem service assessment frameworks developed in the last decade, including the one deployed by the Intergovernmental Science-Policy Platform on Biodiversity and Ecosystem Services (IPBES), a recently established intergovernmental body; and its efforts to inform policy formulation.

What About Natural Capital?

Arguably, one of the most important tasks for today's decision-makers is to develop social and economic governance systems which understand and tackle the effects of environmental change on ecosystems and human well-being. It is at the interface of the ecological and economic spheres that 'nature' is referred to as 'natural capital'. Capital assets include manufactured capital (buildings and machines), human capital (knowledge, skills, experience, and health), social capital (relationships and institutions), and financial capital (monetary wealth), as well as natural capital [2]. Multiple forms of capital interact to generate goods and services. 'Natural capital' refers to the living and non-living components of ecosystems—other than people and what they manufacture—that contribute to the generation of goods and services of value for people [4].

The incorporation of concepts such as 'natural capital' and 'ecosystem services'[1] into decision-making is seen as a step towards addressing the challenge of improving our governance systems [2]. In light of this, the pioneering and influential Millennium Ecosystem Assessment (MA), published in 2005, used a novel conceptual framework for analysing and understanding the effects of policies on ecosystem services and human well-being. However, more than a decade after its publication, we still lack basic information about the dynamics of social-ecological systems and the interrelationship between ecosystem services and human well-being [6].

Natural capital with increased resilience has an enhanced ability to survive, transform, and adapt to change as well as to continue to provide ecosystem services in beneficial ways [2]. The MA drew attention to this and to the importance of natural capital and ecosystem services in supporting human well-being, while highlighting the capacity of social-ecological systems to sustain natural capital and ecosystem services in the face of disturbance and ongoing changes [7]. It brought significant evidence that human actions were leading to alarming declines in the majority of ecosystem services [8] and aimed to catalyse intents and efforts to reverse ecosystem declines.

Ecosystem Services Assessment Frameworks

Since the publication of the MA, there have been a growing number of similar initiatives, which bring together academic communities, civil society, businesses, non-governmental organisations, governments, and international organisations. These institutions have begun incorporating natural capital and ecosystem service into policy and management but it is not yet fully standard practice [2].

[1] Ecosystems sustain human life through 'ecosystem services' (Guerry et al. [2]). Ecosystem services are the conditions and processes of ecosystems that generate—or help generate—benefits for people. These benefits result from the interactions among plants, animals, and microbes in the ecosystem, as well as biotic, abiotic, and human-engineered components of social-ecological systems. Ecosystem services are produced along the full spectrum of heavily managed ecosystems (e.g. agroecosystems) to ecosystems with low human imprint (TEEB [5]).

For example, The Economics of Ecosystems and Biodiversity (TEEB), an influential international project published in 2010, focused heavily on the economic use and evaluation of ecosystems. TEEB aimed at evaluating in monetary terms the contribution of biodiversity and ecosystems within the framework of productive economic systems [9]. The study also aimed at establishing a global standard basis for natural capital accounting, advocating for including the costs of biodiversity and ecosystem loss and degradation in national accounts ([5]).

TEEB represented a global meta-analysis on the economics of nature and drew on numerous case studies to reveal the economic values of biodiversity and ecosystems. It also provided a number of useful entry points for policymakers at various levels (international, national, regional, and local) and businesses to adopt actions towards sustainable development, by linking science and policy at an early stage [9].

Inspired by the international TEEB process, a number of countries started carrying out TEEB studies, or national ecosystem services assessments, such as the UK's National Ecosystem Assessment (NEA) [10], the Spanish NEA [11], or the New Zealand assessment [12]. Other countries followed suit shortly after, such as the Nordic countries, South Africa, Brazil, India, The Netherlands, Germany, and others.

In the European Union (EU), Member States, in line with their contribution to the EU's Biodiversity Strategy, have also begun mapping and assessing the state of the ecosystem and their services in their national territory, with the assistance of the European Commission's tool called Mapping and Assessing of Ecosystems and their Services (MAES). The framework uses the same international classification of ecosystem services developed for environmental accounting purposes as the MA and TEEB [13].

The aforementioned initiatives have been remarkably successful and progressing, popularising and promoting the concepts and use of natural capital and ecosystem services over the past decade, with both public and private sector decision-makers often referring to the importance of these. However, while scientific research has advanced significantly and new institutions, such as IPBES, are still emerging, businesses and governments have not dramatically changed their operations [14].

One of the ongoing challenges is the need to go beyond merely raising awareness of the interdependence of ecosystems and human well-being. There is an unprecedented urgency for advancing the fundamental interdisciplinary science of ecosystem services, and 'implementing this science in decisions to restore natural capital and use it sustainably' [2].

Furthermore, there is in increase in the number of those calling for trends in biodiversity, ecosystem services, and their impact on human well-being, as well as sustainability, to be studied using an integrated approach [15]. While natural capital has played an important role to date in engaging the key stakeholders it has not, as yet, delivered the real change to practice that is demonstrably needed.

The Sustainable Development Goals (SDGs)[2] and Aichi Targets could potentially offer some key elements for this integration and support real change. Although neither 'natural capital' nor 'ecosystem services' are used explicitly to measure progress toward sustainable development, the concepts imply a direct link between biodiversity and human well-being [17]. The Aichi Biodiversity Targets refer to the 20 biodiversity targets established under the Convention of Biological Diversity[3] which place biodiversity into the mainstream decision-making frameworks of policy-makers [18].

Considerable progress has been made in developing integrative frameworks and definitions for ecosystem services and the quantification of indicators (e.g. Kandziora et al. 2013; Maes et al. 2016). However, at this stage, it is unclear to what extent the current state of knowledge in ecosystem service assessments is able to provide the information required for monitoring progress towards the SDGs and the Aichi Targets [6].

Most assessments use a pragmatic approach to select indicators for ecosystem services, often only focusing on those indicators and ecosystem services for which data are readily available [17]. Although this helps to advance the knowledge on ecosystem services in many aspects, it does not necessarily cover the knowledge required to monitor progress towards sustainability. Regions characterised by high vulnerability of ecosystem service supply and human well-being require information on the trends in all aspects of ecosystem services flows, including the

[2] In 2015, the United Nations (UN) adopted the SDGs as the global development agenda seeking to eradicate poverty and ensure environmental sustainability by 2030. The framework includes 17 goals and comprises 169 individual targets, which focus holistically on environmental, social, and economic pillars of sustainable development [16].

[3] https://www.cbd.int/sp/targets.

impact of governance interventions and pressures on social-ecological systems [19].

The IPBES Conceptual Framework: From 'Ecosystem Services' to 'Nature's Contributions to People'

The IPBES represents a joint effort by governments, academia, and civil society to assess, advance, and promote the knowledge about Earth's biodiversity and ecosystem services and the ways in which they contribute to human society, in order to inform policy formulation. IPBES uses a conceptual framework inspired by the MA and the ecosystem service concept popularised by it [6].

One key novelty in the conceptual framework used by IPBES is the notion of 'nature's contribution to people' (NCP). NCPs 'are all the contributions, both positive and negative, of living nature (diversity of organisms, ecosystems, and their associated ecological and evolutionary processes) to people's quality of life' [20]. Beneficial contributions include food provision, water purification, and artistic inspiration, whereas detrimental contributions include disease transmission and predation that damage people or their assets [20]. Many NCPs may be perceived as both benefits and detriments depending on the cultural, socioeconomic, temporal, or spatial context [6].

The use of NCPs sets apart the IPBES conceptual framework from any other conceptual framework in a number of distinct ways. The NCP approach not only recognises the central role that culture plays in defining the links between people and nature [21], but it also greatly emphasises and operationalises the role of indigenous and local knowledge in understanding nature's contribution to people [6].

The broad remit of IPBES requires it to engage a wide range of stakeholders, spanning from natural, social, humanistic, and engineering sciences to indigenous peoples and local communities in whose territories lie much of the world's biodiversity [21]. Being an intergovernmental body, such inclusiveness is essential not only for advancing knowledge but also for the political legitimacy of assessment findings [20].

There is a marked difference in approach within IPBES from the earlier MA and MA-inspired initiatives, as they were dominated by knowledge drawn from the natural sciences and economics [21]. The natural sciences, and ecology in particular, were used to define 'ecological production functions' to determine the supply of services, conceptualised as flows stemming from ecosystems (stocks of natural capital). Economics was used to estimate the monetary value of those ecosystem service flows so as to identify trade-offs among them and their impacts on well-being [22].

This predominantly 'stock-and flow' framing of people-nature relationships largely failed to engage a range of perspectives from the social sciences [23], or those of local practitioners, including indigenous peoples. This reinforced a mutual alienation process in which the MA, and much of the studies and policies which were inspired by it later on, became increasingly narrow, which in turn led to voluntary self-exclusion of disciplines, stakeholders, and world views [21].

As a consequence, the ecosystem services research programme proceeded largely without benefiting from insights and tools in social sciences and humanities [23]. The need to be inclusive, both in terms of the strands of knowledge incorporated and representation of world views, interests, and values, required IPBES to move to using NCP [6].

Although still rooted in the MA ecosystem services framework, this new approach aims to firmly embed and welcome a wider set of viewpoints and stakeholders, including diverse and less represented knowledge systems [23]. It also aspires to be less likely integrated within a narrow economic (such as market-based) approach as the mediating factor between people and nature [24].

With the NCP framework firmly embedded in the IPBES processes, a number of regional and thematic, and one global, assessments are planned for 2018 and 2019. These involve hundreds of scientists, analysing various aspects of human-nature interactions, the drivers of biodiversity loss, and strategies for linking people and nature within models [20].

Critical Response to NCP

A paper published in *Science* authored by Diaz et al. [6] which includes contributions from a number of scientists associated with IPBES argued that the NCP framework could represent a 'paradigm shift from the concept of ecosystem services'. The paper elicited a number of critical responses in peer-reviewed journals such as *Ecosystems Services* and *Science*. The criticisms were primarily levelled at the justification of producing an actual 'paradigm shift' in terms of scientific advancement in the field of ecosystem services [25].

A number of follow-up responses to the Diaz et al. [6] paper have positively pointed out that the NCP concept places the cultural context (and the way in which it contributes to 'human perception of nature and good quality of life') into a very sharp focus [24]. However, by replacing 'ecosystems' with 'nature', the NCP concept may have inadvertently underemphasised the extent to which 'social-ecological processes shape the world's ecosystems' and how much of the 'Earth's systems are shaped by human action' [24].

NCPs may also de-emphasise other aspects of the human-nature relationship which include but are not limited to, for example, the feedback between ecosystems and how benefits from them are derived; or the wider role that technology and infrastructure play in producing or accessing ecosystem services [24]. As such, NCPs ignore the complexity of co-creation and co-production of ecosystem services by emphasising a one-directional flow from 'nature' to 'people' [25].

However, despite its limitations, NCP has a better potential for attracting and involving a wider range of stakeholders and policymakers than the ecosystem services concept [6]. This should not automatically discount the tremendous policy advancements and achievements ecosystem services have already achieved in communicating the value of ecosystems and biodiversity, like some critics have claimed [25]. It is equally true though that there is simply not enough substantial evidence to prove that the NCP concept will be more useful in policy terms (or indeed create a paradigm shift) than ecosystem services [24].

In our view, the Diaz et al. [6] paper, and indeed the IPBES conceptual framework itself, does not propose replacing ecosystem services with NCPs. We do not recognise the need for the level of polarisation this debate has sparked in the scientific community. Rather, much like Peterson et al. [24], we argue that practical knowledge for sustainability requires diverse approaches and different ways of thinking.

Furthermore, as pointed out by Diaz et al. [6], avoiding division enables greater collaboration and invites the kind of scientific pluralism ecosystem services research and policy urgently needs [6]. To achieve this, IPBES, through its structure and mandate, could do more in working together more broadly and inclusively within already existing global networks of scientific programmes and environmental assessments [24].

Natural capital—nature—will be destroyed unless policy and economic activity can recognise the coupling between economy and environment. Whether we like it or not, our human-nature relationship is now fundamentally a two-way process and we must deploy all possible routes to build a better understanding within those communities who hold key influence over how this relationship changes with time.

IPBES represents one of these narratives and we believe it can be widely deployed to contribute to this understanding. However, as with a number of other approaches to natural capital, it cannot operate alone or we would risk further damage to nature—especially to that part of nature which is not captured in the specific boundaries of the NCP approach adopted through the remit of IPBES.

References

1. Ring, I., B. Hansjürgens, T. Elmqvist, H. Wittmer, and P. Sukhdev. 2010. Challenges in framing the economics of ecosystems and biodiversity: The TEEB initiative. *Current Opinion in Environmental Sustainability* 2: 15–26.
2. Guerry, A., S. Polasky, J. Lubchenco, R. Chaplin-Kramer, G.C. Daily, R. Griffin, M. Ruckelshaus, I.J. Bateman, A. Duraiappah, T. Elmqvist, T. Feldman, C. Folke, J. Hoekstra, P. Kareiva, L. Keeler, S. Li, E. McKenzie, Z. Ouyang, B. Reyers, T. Ricketts, J. Rockström, H. Tallis, and B. Vira. 2015. Natural capital informing decisions. *Proceedings of the National Academy of Sciences* 112 (24): 7348–7355.
3. Biggs, R., M. Schlüter, D. Biggs, E.L. Bohensky, S. BurnSilver, G. Cundill, V. Dakos, T.M. Daw, L.S. Evans, K. Kotschy, A.M. Leitch, C. Meek, A. Quinlan, C. Raudsepp-Hearne, M.D. Robards, M.L. Schoon, L. Schultz, and C. Paul. 2012. Toward principles for enhancing the resilience of ecosystem services. *West Annual Review of Environment and Resources* 37 (1): 421–448.
4. Schultz, L., C. Folke, H. Österblom, and P. Olsson. 2015. Adaptive governance, ecosystem management, and natural capital. *Proceeding of the National Academy of Sciences USA* 112: 7369–7374.
5. Mainstreaming the Economics of Nature: A Synthesis of the Approach, Conclusions and Recommendations of TEEB. 2010. Available at www.teebweb.org/publication/. Accessed 10 Apr 2018.
6. Diaz, S., U. Pascual, M. Stenseke, B. Martin-Lopez, R. Watson, Zs Molnar, R. Hill, K.M.A. Chan, I. Baste, K. Brauman, S. Polasky, A. Church, M. Lonsdale, A. Lariguardie, P. Leadley, P.E. Van Oudenhoven, F. Van der Plaat, M. Schroter, S. Lavorel, Y. Aumeeruddy-Thomas, E. Bukvareva, K. Davies, S. Demissew, G. Erpul, P. Failler, C. Guerra, C. Hewitt, H. Keune, S. Lindley, and Y. Shirayama. 2018. Assessing nature's contributions to people. *Science* 359: 270–272.

7. Folke, C., S.R. Carpenter, B. Walker, M. Scheffer, T. Chapin, and J. Rockström. 2010. Resilience thinking: Integrating resilience, adaptability and transformability. *Ecology and Society* 15 (4): 20.
8. Scheffer, M., S. Barrett, S.R. Carpenter, C. Folke, A.J. Green, M. Holmgren, T.P. Hughes, S. Kosten, I.A. van de Leemput, D.C. Nepstad, E.H. van Nes, E.T.H.M. Peeters, and B. Walker. 2015. Climate and conservation. Creating a safe operating space for iconic ecosystems. *Science* 347 (6228): 1317–1319.
9. Ulgiati, S., A. Zucaro, and P. Franzese. 2011. Shared wealth or nobody's land? The worth of natural capital and ecosystem services. *Ecological Economics* 70: 778–787. https://doi.org/10.1016/j.ecolecon.2010.11.015
10. The UK National Ecosystem Assessment. 2011. *The UK National ecosystem assessment: Synthesis of the key findings*. Cambridge: UNEP-WCMC.
11. Santos-Martín, F., B. Martín-López, M. García-Llorente, M. Aguado, J. Benayas, and C. Montes. 2013. Unraveling the relationships between ecosystems and human wellbeing in Spain. *PLoS One* 8: e73249.
12. Dymond, J. 2013. *Ecosystem services in New Zealand*. Lincoln: Landcare Research New Zealand, Manaaki Whenua Press.
13. Biodiversity Information System for Europe. 2018. https://biodiversity.europa.eu/maes. Accessed 10 Apr 2018.
14. Vohland, K., and T. Nadim. 2015. Ensuring the success of IPBES: Between interface, market place and parliament. *Philosophical Transactions of the Royal Society, B: Biological Sciences* 370 (1662): 20140012.
15. Bennett, E., W. Cramer, A. Begossi, G. Cundill, S. Diaz, B. Egoh, I.R. Geijzendorffer, C. Krug, S. Lavorel, E. Lazos, L. Lebel, B. Martín-López, P. Meyfroidt, A. Mooney, Unai Pascual, K. Payet, P. Harguindeguy, G. Peterson, and G. Woodward. 2015. Linking biodiversity, ecosystem services, and human well-being: Three challenges for designing research for sustainability. *Current Opinion in Environmental Sustainability* 14: 76–85. https://doi.org/10.1016/j.cosust.2015.03.007
16. Josephsen, L. 2017. Approaches to the implementation of the sustainable development goals – Some considerations on the theoretical underpinnings of the 2030 Agenda. Economics Discussion Papers, No 2017–60, Kiel Institute for the World Economy. http://www.economics-ejournal.org/economics/discussionpapers/2017-60.
17. Geijzendorffer, I.R., E. Cohen-Shacham, A.F. Cord, W. Cramer, C. Guerra, and B. Martín-López. 2017. Ecosystem services in global sustainability policies. *Environmental Science and Policy* 74: 40–48.
18. Rode, J., H. Wittmer, and G. Watfe. 2012. Implementation guide for Aichi Target 2 – A TEEB perspective. German Federal Agency for Nature Conservation (BfN).
19. Hill, R., G.A. Dyer, L.M. Lozada-Ellison, A. Gimona, J. Martin-Ortega, J. Munoz-Rojas, and I.J. Gordon. 2015. A social–ecological systems analysis

of impediments to delivery of the Aichi 2020 Targets and potentially more effective pathways to the conservation of biodiversity. *Global Environmental Change* 34: 22–34 ISSN: 0959–3780.
20. IPBES Plenary 5 Decision IPBES-5/1. 2015. Implementation of the first work programme of the platform, p. 23. www.ipbes.net/event/ipbes-5-plenary.
21. Diaz, S., S. Demmisew, J. Carabias, J. Carlos, M. Lonsdale, N. Ash, A. Lariguade, J.R. Adhikari, S. Arico, A. Baldi, A. Bartuska, I.A. Baste, A. Bilgin, E. Brondizio, K.M.A. Chan, E. Figureroa, A. Duraiappah, M. Fischer, and D. Zlanatova. 2015. The IPBES conceptual framework—Connecting nature and people. *Current Opinion in Environmental Sustainability* 14: 1–16.
22. Millenium Ecosystem Assessment Panel. 2005. *Ecosystems and human well-being: Synthesis*, Millenium ecosystem assessment series. Washington, DC: Island Press.
23. Maes, J., B. Burkhard, and D. Geneletti. 2018. Ecosystem services are inclusive and deliver multiple values. A comment on the concept of nature's contributions to people. *One Ecosystem* 3: e24720.
24. Peterson, G.D., Z.V. Harmackova, M. Meacham, C. Queiroz, A. Jiménez Aceituno, J.J. Kuiper, K. Malmborg, N.E. Sitas, and E.M. Bennett. 2018. Welcoming different perspectives in IPBES: "Nature's contributions to people" and "Ecosystem services". *Ecology and Society* 23 (1): 39.
25. Braat, Leon C. 2018. Five reasons why the science publication assessing nature's contributions to people (Diaz et al. 2018) would not have been accepted in ecosystem services. *Ecosystem Services* 30 (A): A1–A2. https://doi.org/10.1016/j.ecoser.2018.02.002.

CHAPTER 3

A Micro 'Case Study': Critiquing the Economics of Ecosystems and Biodiversity

Rupert Read and Tom Greaves

Abstract In this short case study we critically assess The Economics of Ecosystems and Biodiversity (TEEB) framework in the light of an ecological-relational view of the source of value. This view takes all value to be generated by the whole range of relations between living beings. We show that whilst the TEEB has advantages over some rival frameworks, allowing limited room to assign value to the non-human, it still does not take seriously enough the participation of non-human life in the very processes of value formation.

In the following couple of pages, we briefly outline a critique of the increasingly influential 'The Economics of Ecosystems and Biodiversity' (TEEB) approach to valuing life, which is paradigmatically representative of the trend of 'economic valuation of ecosystem services'. We think that

Excerpted [with a very light edit] from our paper, "Where values reside", *Environmental Ethics* 37 (2015): https://www.researchgate.net/publication/290473224_Where_Value_Resides_Making_Ecological_Value_Possible

R. Read (✉) • T. Greaves
School of Politics, Philosophy, Language and Communication Studies, University of East Anglia, Norwich, UK
e-mail: R.Read@uea.ac.uk

V. Anderson (ed.), *Debating Nature's Value*,
https://doi.org/10.1007/978-3-319-99244-0_3

the way we have sought to re-present valuing (in our paper "Where value resides"[1] sheds light on some damaging tacit philosophical presumptions in the mainstream discourse now accumulating around 'biodiversity'. We'll show how, here.

We take as our text the influential 'Teeb for policy-makers' document.[2]

Let us begin with a couple of early warning signs. We are told up front in this document that "around one European job in every six is somehow dependent on the environment" (p. 24). An adequate understanding of the word 'environment', especially when the claim is qualified by 'somehow', suggests that the figure should actually be six jobs in every six. The slip is revealing of the extent to which this 'ecosystem-based' approach is still tacitly figuring the Earth as a resource and an 'externality' to the 'real' economy. Its being and its value are then bound to be instrumentally conceived, not something to which we are internally *related*. To sum up this point: conceiving of nature as external to us is a bad mistake.

The TEEB regards "biological resources [as] a stock of capital in their own right" (p. 14). But this is still problematic. Using 'capital' as one's founding metaphor already implicitly figures nature as something to be placed under an economic system of measure. Biological resources become a *kind* of capital. We assume that we know everything that there is to know about capital, and that we can then carry that over directly to refigure nature. Nature is then placed essentially under the 'law-table' of human beings and our economics. Any sense of otherness, or of 'intrinsicness', or even of betweenness, is already over once this move has been made. To sum up this point: thinking of nature as 'internal' *to us*—that is, internalising it within the human economy—risks being just as bad, just as deep a mistake as thinking of it as 'external'. It risks figuring ecology as just *another extended version* of that economy. The mistakes of externalisation and internalisation display the same failure as thinking of an ecological *community of difference* where valuation takes place between the different members.

[1] Full text available here: https://www.researchgate.net/publication/290473224_Where_Value_Resides_Making_Ecological_Value_Possible

[2] Wessseling: Welzel and Hardt, 2009; http://www.teebweb.org/Portals/25/Documents/TEEB%20for%20National%20Policy%20Makers/TEEB%20for%20Policy%20exec%20English.pdf

Here is a particularly striking example of the consequences. In the course of a discussion of the economic value of coral reefs, we read the following: "Coral reefs are now understood to have a critical range of ecosystem service values – for natural hazard management, ... tourism, ... genetic materials and bio-prospecting, ... fisheries. ... These benefits are site-specific – so a global loss of coral reefs will impact communities differently. *Lost benefits will be lowest in places with few people,* poor ecosystem quality *or limited accessibility. ... The lowest values generally correspond to sites with limited accessibility or facilities for tourism, while the very high values relate to international tourism hotspots*" (p. 11). Coral reefs with limited accessibility are less valuable than those on the tourist trail. There is no room here, it seems, for the idea that we might rightly value coral reefs that are inaccessible to us *more* than those that are accessible. That the less accessible reefs might, through being less accessible, be more valuable and/or valuable in quite a different way. If I value some remote part of the Great Barrier Reef or the Queensland rainforest, that does not necessarily mean that I will want to be in that reef or forest or as close to it as possible, for as much of my life as possible. It may be that I value it *by* not going into it *at all.* I might value it precisely as wild nature. *It* may even be of value *through* not being part of any active focused present prizing by me or anyone else. (In other words, I might well value precisely that there are things that I do not put any first-order value on. We can escape the seeming limit of whatever we value seemingly being of value only in virtue of being prized by us.) So long as it is encompassed by the co-constituting worlds of living beings, feeding in some way or another into their potential, it can be valuable even if none is aware of that value or even if it is not the kind of thing that they could become aware of as individuals.

Of course, the very concept of 'ecosystem services', central to TEEB, and explicit in the quotation given earlier, is itself deeply troubling. It figures the planet as a source of services *to us*, the ultimate 'service provider'. One that does not even need to be paid for, since any money that changes hands necessarily goes to humans: to governments, corporations, and so on. Once one questions the *concept* of 'ecosystem services', one has to question the very idea of TEEB.

Similar concerns could be raised with regard to other currently influential projects for 'valuing biodiversity'. For example, there are clearly things to welcome in the massive, newly publicised Natural Environment Research Council (NERC) project "Valuing Nature" (www.valuing-nature.net), including the remarks of its Principal Investigator, our former colleague here at University of East Anglia (UEA), Ian Bateman, to the effect that the value of nature is, strictly speaking, infinite.[3] What Bateman says on this score is far less troubling than the (nowadays ubiquitous[4]) claim of Costanza et al., that a price tag can notionally be put on the *whole* of nature, in terms of the 'ecosystem services' it delivers.

Yet it is troubling that Bateman nevertheless refuses to consider the possibility of nature having (as he puts it) 'intrinsic value'. He defends this refusal with the argument that we cannot know if we are valuing nature as it itself (or they themselves, in the case of non-human animals) would want to be valued.[5] How do we know what a whale values, Bateman asks, rhetorically? This is somewhat ironic as a dogmatic assertion (that we do not, and so cannot, ascribe them any 'intrinsic' value) made in the course

[3] See http://www.ft.com/cms/s/2/644b1884-8c7c-11e0-883f-00144feab49a.html#axzz1O6Gtm2d6 & http://www.bbc.co.uk/news/science-environment-13616543. Bateman's stance is thus better than the available alternatives, to date. But we want to look towards something post-NERC (and post-TEEB) that would actually be good enough. 'Intrinsic value' will always lose in the balance if it is the intrinsic value of, for example, a forest, something non-human. Instrumentalist CBA must be rejected/transcended, because of incommensurability (because of the reality of the infinite value of life, of the whole of nature (though our perspective in the present chapter *also* puts into question whether it can really *make sense* to 'sum' the whole of nature, at all)). We are trying to offer the schematic beginning of a viable alternative that could find its way eventually into a policy document that would transcend the TEEB (and the NERC work of Bateman et al.), an alternative that would be based on 'placing' *value* itself *between*. And that would thus overcome the dubious meaning that the term 'environment' has come to have, of something 'out there' to be 'managed', and would midwife a recovery of the earlier, better, encompassing, more 'ecological' sense of the term 'environment', a sense that is to some degree already afforded by the gap-dissolving concept of 'affordances'.

[4] See https://www.ons.gov.uk/economy/environmentalaccounts/bulletins/uknaturalcapital/monetaryestimates2016 for an up-to-date egregious example from the UK.

[5] Bateman's argument (presented in ENV, UEA, on 31 March 2011) is reminiscent of the argument currently, tragically, being used to justify the destruction of the world's greatest seed bank, which in older, perhaps better, days scientists died of starving from rather than consuming (see http://www.care2.com/causes/real-food/blog/historic-russian-seed-bank-faces-destruction/) in Moscow: "Under [a] mind-boggling quirk, the fact that the land is deemed as 'priceless' also gives developers a right to build there. To legally prove that we are using these lands, we need to put a value on the collection which is impossible. There are no methods for that. How can we put a price on a collection that is unique and only exists here" (http://www.expatica.ru/news/news_focus/Outrage-as-priceless-Russian-seed-bank-faces-destruction_92206.html). 'Priceless' becomes equated with valueless. This is an obscene turning on its head of our system of value.

of an investigation into the value of nature, since when it comes to assessing the values held dear by human beings we are frequently reminded that behaviour is usually a far better indicator of 'preference' than the explicit claims that people make. In other words, advocates of cost-benefit analysis (CBA) and/or randomised control trial (RCT) have particularly little ground to stand on, in resisting the idea that the preferences/values of non-human animals need to be taken seriously. Similarly, we can *look and see* what a whale values. Their preferences are revealed by their actions.[6]

It is particularly ironic that Bateman picks on whales—for whales are surely not merely open, but manifest some open-ended openness. In their capacity, for instance, for communicative song, song that exhibits development and regional/'tribal'/personal variation, they appear to be, subtly, growing ecological valuers. We humans have still a great deal to learn about whales; they *sound* in love with each other, and/or with the(ir) world.

And, even if this were not the case, Bateman's rhetorical question is in any case making the classic move that we diagnosed and critiqued earlier in the chapter. He moves smoothly from something not necessarily being valued *by 'us'* to its not being of value. Whereas value requires ecological valuers as its condition, it is by no means exhausted by that condition.

If our argument is right, the truth that trees do not suffer as we do and the (far smaller) element of truth there might be in the claim that there is something potentially problematic in ascriptions of interests to whales (for we submit that difficulties in understanding what exactly it is like to be a whale do not gainsay that whales clearly have not only vital but also cognitive interests, and that they join us in (helping to constitute!) a community of what we call 'open-endedly open beings') does not in any case in any way licence Bateman's refusal to countenance intrinsic value. Rather, we need to be open to the different *kinds* of value that are to be found in living alongside and amongst living beings other than ourselves. Just because this value cannot be assessed in the terms that one is used to or even in further broadenings of those terms (the kind of broadening found, for instance, in Costanzian 'ecological' economics, which the TEEB and NERC projects are inheritors of) does *not* imply that it simply cannot or should not be talked about or taken seriously. On the contrary, it gives us

[6] Moreover, ecological value is not just a function of the sum of all revealed preferences, but the outcome of complex interrelations and encounters *between* those preferring.

all the greater *responsibility* for taking it seriously. We need to ensure precisely that there is space for the value of what and where we are not. Value lies between us, between us, and where we are not. It cannot and must not be pulled back onto and into us alone. This is true, we have suggested, *regardless* of the extent to which we understand other beings.

Perhaps Bateman and others thinking similarly to him are caught up in the allegedly stark opposition of intrinsic to instrumental value. Our suggestion is that the old saw of environmental ethics, 'intrinsic value', is better understood as or changed into a *between* than as inhering solely and fully *in* other things (i.e. we need to transcend the dichotomy of nature as either 'external' to us or 'internal' to us). Every 'instrumental' value requires something 'intrinsic' to 'measure' it by and every intrinsic value that we are open to must be appraised in practice. Our own value is *inextricable* from the value of that around us. The valuations of non-human animals are values that are to some significant extent revealed by *their* preferences, but are also inextricable from our own values, once we understand the latter broadly enough. The value of those living things that are presumably not at all open-endedly open (trees, microbes, etc.) is an event involving them and us and non-human animals. The value of the non-living depends upon the valuations of open beings, and the value of both the former and the latter is enriched further by the depth or fullness of the world—of the community—made possible by the existence of open-ended openness (Heideggerian Dasein or something akin to it, relatively widespread and broadly reciprocal).

For these reasons, then, we believe that philosophy has a vital role to play in the project of valuing nature. There is a grave danger in the fact that at present the 'Valuing Nature' project and the 'TEEB' project do not contain explicit philosophising. The danger is that they will be dominated by an implicit philosophy, by an outdated merely humanist ethic that has a rampantly insufficient appreciation even of our humanity. Human beings value more than is dreamt of in this implicit philosophy. Philosophy is needed if such projects are to take account of the ecological conditions of valuation and to develop and deploy a properly ecological concept of value. Potential for doing so can be found in the 'Valuing Nature' project, which does recognise the non-instrumental value of biodiversity and healthy ecosystems. But it tends to overlook or inadequately address the ecological conditions for such valuation. Once CBA of various projected futures has been carried out, the approach theoretically allows various kinds of valuation to be brought to bear so as to *veto* policies leading to some scenarios, even if they are the 'best' scenarios according to the analysis. This is a welcome step in the right direction. However, the attempt to preserve the purely quantitative analysis from such 'external' cares, concerns, and values remains troubling. Rather than the blunt instrument of veto after the fact, such 'externalities' ought to be internalised into the very notion of the valuations taking place. We need to find a way of sufficiently acknowledging in our policy processes the valuableness of beings and places that are truly beyond price.

The economics of ecological valuation needs to move beyond first appearances and initial desires that have been bound to ideals of functionality, instrumentality, and homogenising anthropomorphism. Collectively, we *must* provide some genuine space for the critical 'places'—the non-*place*—where value truly resides. We must somehow find a way of leaving enough space for the valuable that is not merely human, all-too-human.

CHAPTER 4

The Natural Capital Protocol

Samir Whitaker

Abstract Current decision-making processes do not adequately account for biodiversity and ecosystem services, apparent in the downward trends for many global indicators of environmental health. Globally, close to half of the largest economies are companies, and an increasing awareness of how businesses are degrading ecosystems has led to a plethora of new methodologies to value biodiversity and ecosystem services as well as increasing consumer and investor pressure to better account for and manage these risks. Resistance to this pressure is waning as CEOs are seeing the value proposition: opportunities to secure supply chains, differentiate from competitors with new products, and respond to shareholder calls for more holistic risk assessment and management. With overlapping methodologies, frameworks, and guidance forming the beginnings of a crowded and potentially confusing landscape, the Natural Capital Protocol aims to serve as 'the' standardised framework for businesses to measure and value natural capital, providing results to guide corporate decision-making. In this chapter, I explore the history and need for the Protocol, its structure and development, and some recent criticism of Natural Capital approaches.

S. Whitaker (✉)
Zoological Society of London, London, UK
e-mail: Samir.Whitaker@zsl.org

V. Anderson (ed.), *Debating Nature's Value*,
https://doi.org/10.1007/978-3-319-99244-0_4

Mainstream decision-making processes do not adequately account for biodiversity and ecosystem services because much of it, biodiversity in particular, does not readily convert into economic terms and requires, therefore, repackaging. With growing interest from the investment community in more holistic understanding and management of risk, and from governments in policy to secure and promote the wise use of these resources, natural capital is fast becoming a mainstream concept [1]. From a policy perspective, the first formal European Union (EU) rules on environmental accounting were established in 2011, with the Regulation on European Environmental Economic Accounts [2]. This rule requires EU Member States to regularly report on three areas to the European Statistical Office (Eurostat), including accounts on environmental taxes and economy-wide material flow accounts. Globally, the Convention on Biological Diversity's Strategic Plan for Biodiversity 2011–2020, containing the 20 'Aichi Biodiversity Targets, the second of which is that '[b]y 2020, at the latest, biodiversity values have been integrated into national and local development and poverty reduction strategies and planning processes and are being incorporated into national accounting, as appropriate, and reporting systems' [3].

Increasing consumer, investor, and government pressure to understand, account for, and internalise natural capital dependencies is driving businesses that are not already exploring these concepts to begin to do so. An increasing awareness of how we are degrading ecosystems and the need for tools and approaches has led to a plethora of new frameworks and methodologies to value ecosystem services—the flow of benefits from natural ecosystems, and natural capital—the stock from where these benefits are derived. These tools and approaches include the World Bank's Wealth Accounting and the Valuation of Ecosystem Services (WAVES) partnership, Cambridge Conservation Initiative's Toolkit for Ecosystem Services Assessment (TESSA), World Business Council for Sustainable Development's (WBCSD's) Business Ecosystems Training (BET), TEEB for Business, and the European Commission's Knowledge Innovation Project on Accounting for Natural Capital and Ecosystem Services (KIP INCA) [4].

With overlapping methodologies, aims, and guidance forming the beginnings of a crowded and potentially confusing landscape, the Natural Capital Protocol (NCP) was formulated in order to serve as a consolidated guidance for businesses. According to the Natural Capital Coalition (NCC), '[t]he Natural Capital Protocol is a framework designed to help generate trusted, credible, and actionable information for business managers to inform decisions. The Protocol aims to support better decisions by including how we interact with nature or more specifically natural capital' [5], and builds on a number of approaches that already exist to help business measure and value natural capital, including the Corporate Ecosystem Services Review [6], and the Guide to Corporate Ecosystem Valuation [7].

The Protocol is designed for use by the business sector, and aims to be a standardised resource to guide business users through the identification and valuation of natural capital relevant to that business. As it is primarily designed to guide internal reporting and decision-making, and does not prescribe a standard methodology for valuation, its utility as an external reporting tool is limited as it does not have the ability to benchmark information produced by businesses using the protocol, unless they use comparable methodologies. As the Protocol is focused on internal decision-making, and it is not a reporting tool; however, it can be used for this purpose with the caveats listed above.

A Potted History

> Fossil fuels are merely a part of the 'natural capital' which we steadfastly insist on treating as expendable, as if it were income, and by no means the most important part. If we squander our fossil fuels, we threaten civilisation; but if we squander the capital represented by living nature around us, we threaten life itself. (Ernst F. Schumacher) [8]

This passage, from Schumacher's book *Small Is Beautiful: A Study of Economics as if People Mattered* [9] is one early reference to the term 'natural capital'. Also prescient, or what we would now write off as 'common sense', is Schumacher's recognition and overt description of natural capital as the basis for the benefit flows that sustain human life. For the period this was revolutionary thinking, and considering his training as an economist and role as Chief Economic Adviser to Britain's National Coal Board it is even more remarkable. Perhaps it does make sense, however, working as he was with a non-renewable resource that was already known to have far-reaching consequences for human and environmental health. Over a decade later, David W. Pearce was, according to Missemer [10], the first economist to use the expression 'natural capital' to refer to natural resources and services, calling them 'the set of all environmental assets'. This was some of the early thinking underscoring the need to account for the stocks from which vital services flow, and to include in global economic accounts.

Understanding the need for integrated national accounting began in 1930s USA when economists were struggling to catalogue and understand the full scale of The Great Depression, one of the vital data gaps being the lack of consolidated and comprehensive data on national income and outputs; America did not understand how much it was selling and how much it was earning. The resulting system of national accounts created to address this data gap was so vital to policymakers that a standardised system was then developed by the United Nations (the UN System of National Accounts), and was used by member states to report back on national incomes, deficits, and savings and became a crucial and widely used tool for policy formulation and tracking.

Fifty years later, initially driven by risks associated to the availability of water in drought-prone regions, several countries—Australia as the leader, for obvious reasons—began including environmental factors as part of their System of National Accounts (SNA). Including factors such as the key users and pricing of water 'helped ensure that the most critical and efficient users got water'. Building on these initial indicators, Australia has now expanded its suite of natural capital accounts to include energy, minerals, land, and environmental protection expenditure. Recognising the need to account for underlying systems and resources on which national incomes depend, such as forest and agricultural land and mineral resources, the World Bank began 'a global database for comprehensive wealth accounts in the 1990s. This led to the development of the associated macroeconomic indicator of Adjusted Net Savings (ANS), which measures whether the growth a country is experiencing, adjusted against how it is using its natural capital, is sustainable [11].

Another key milestone was the Millenium Ecosystem Assessment, called for by then UN Secretary-General Kofi Annan, published in 2005. One of the main aims of the assessment was to explore how ecosystem change could affect human well-being, and it was the first comprehensive global audit of natural capital. The results of the assessment were stark: 60% of the 24 ecosystem services examined were being degraded, and their underlying capital being unsustainably exploited. The headline finding of the study was that the 'ability of the planet's ecosystems to sustain future generations can no longer be taken for granted' [12].

The precursor of the NCC, which produced the NCP, was The Economics of Ecosystems and Biodiversity (TEEB), a global study launched in 2007 to explore the contributions of biodiversity and ecosystem services (BES) to the global economy, and simultaneously to explore the risks of degrading these goods and their underlying systems. This was, in their own words, 'a global initiative focused on making nature's values visible', the overarching goal of the study being to encourage the mainstreaming of BES into traditional accounting systems and thereby 'ensure efficient use of resources for which prices are lacking' [13].

Building on growing global recognition of the importance of BES issues, driven significantly by the release of the TEEB study, the TEEB for Business coalition formed in 2012 to 'take action and achieve tangible outcomes' [14]. Recognition was growing of the fact that many of the benefits of BES were at best undervalued, if reflected at all, in the market price of global goods and services and that this must be addressed to avoid further degradation of these systems. A 2013 analysis by the consultancy TruCost, Natural Capital at Risk—The Top 100 Externalities of Business, estimated that the total annual cost of environmental externalities, for example loss of environmental services such as carbon storage and air pollution-related health costs, was $4.7 trillion, at a time when global Gross Domestic Product (GDP) stood at just over $76.5 trillion [15].

In June 2012, the Rio+20 conference 'marked a watershed in the world wide interest on Natural Capital Accounting' [16]. The World Bank started the 50:50 initiative inviting the public and private sectors to join forces and take collective action to support natural capital accounting. By the time of the conference, 62 countries, 90 corporations, and 17 civil society members had joined the campaign and just before the conference, The Natural Capital Declaration (NCD) was announced. This is a declaration from the financial sector to 'embed environmental, social and governance (ESG) considerations in management and investment activities. The

NCD was signed by the CEOs of more than 40 financial institutions, demonstrating their commitment to integrating natural capital considerations into private sector reporting, accounting and decision-making by 2020' [17].

In 2014, with global momentum building for the mainstreaming of natural capital into business and national accounts, but with the recognition that business would not be able to tackle these issues alone, the initiative previously known as TEEB for Business was rebranded as the NCC. The coalition is currently hosted by the Institute of Chartered Accountants in England and Wales (ICAEW), which provides the legal and financial operating framework for the coalition's Secretariat to operate. The NCC, working with 40 businesses initially, developed the NCP, and a subset of ten businesses tested it against various applications specific to their business [18].

Framing the Protocol

The NCP, approaching its two-year anniversary, has gained a great deal of traction as the go-to resource for national capital assessments. In order to remain applicable to a broad range of industries and geographies, however, the protocol does not recommend a specific valuation methodology; it instead contains an overview for a broad range of methodologies. Although it may be used by governments and other institutions, the protocol is primarily aimed at business and that is how the modules are set up.

The Natural Capital Protocol is a framework for a company planning to conduct an assessment of its natural capital dependencies. The flow of the protocol is broken up into four broad stages: the framing, which details why an assessment is being done; the scope or target; the measure and value stage, which is where the chosen methodology is applied; and the apply or internal decision-making stage (Table 4.1).

The stages outlined in the protocol are designed to be iterative, allowing for the original scope to be revisited to ensure that a chosen methodology will produce results that meet the original requirements for example, or going back to include a previously unrecorded dependency or opportunity.

Table 4.1 Understanding and applying the protocol

Frame	*Scope*	*Measure and value*	*Apply*
Understand the end use of the assessment, as this will dictate the approach, resource allocation, and time frames. This stage considers which dependencies might be relevant and how the information might benefit the business.	This stage confirms the focus of your assessment—the natural capital impacts most relevant to the business and its stake (or rather stock) holders. To ensure a consistent assessment the defined scope should be followed.	Consistency of approach is crucial to ensure comparability. This stage involves applying a chosen methodology to measure impact drivers, value natural capital, or understand trends following the scope.	Interpret and verify results before they are used for decision-making processes. It is crucial to regularly check that the results are relevant to the framing and scoping steps.
←	Stakeholder consultation Consistency of approach Document decisions		→

Driving Uptake

The NCC is relying on its members and global partners to drive uptake and support implementation of the protocol, and is now focused on its endorsement by governments and uptake by businesses globally. This approach involves the creation of the Natural Capital Hub through the Coalition's website, which is a resource for developments in the field of natural capital, including case studies of companies using the Protocol. Part of this effort is the creation and release of specific sector guides working with partners such as WBCSD and the consultancy firm TruCost. Sector guides have now been produced for the apparel sector, the food and beverage sector with others, such as forestry, currently under discussion and/or in draft form. A supplement for financial institutions has also been recently developed with the Natural Capital Finance Alliance [19], which provides a framework to assess impacts and dependencies of investment or insurance portfolios. The launch of the finance sector guidance was supported by a series of case studies from financial institutions such as BNP Paribas Asset Management, Kepler Chevreux, and YES Bank.

The sector guides provide more specific hypothetical examples, guidance, and justification for the application of the protocol to a specific sector as well as further background information. The apparel sector guide, for example, includes discussion of the value chain and provides a sector-specific risk and opportunity summary. The guides also drive uptake, as businesses respond to signals from their peers and competitors. To drive further uptake, the NCC also set up and drove the NCP Application Programme, run on behalf of the coalition by the University of Cambridge Institute for Sustainability Leadership, which supported 100 businesses in 40 countries to apply the protocol and sector guides. It also engaged with a similar number of trainers and technical advisers, and maintains a public database of these advisers as organisations that 'intend to encourage the use of the Natural Capital Protocol as a global standardised framework for understanding impacts and dependencies on natural capital and integrating this into decision making' [20]. Several of the businesses involved in the programme such as Jaguar Land Rover, Dow Chemical, Roche, and Tata Group have agreed to publicly release case studies summarising why they chose to do an assessment, which methodology they used, and what the outcomes were. Yorkshire Water, for example, used the NCP to guide their decision on options to replace an existing water filtration plant. Using an agreed subset of valuation methodologies from peer-reviewed literature they examined projected Net Present Value (NPV) for each option across a range of four ecosystem services: climate regulation, air quality, pollination, and cultural and spiritual services [21]. Although relative NPV in this case was really comparing just two options, the case study does serve as a useful guide to business as to why they might undertake such an assessment.

Perhaps understandably, given the focus of the NCP, government engagement appears to be lagging somewhat, with a recent analysis [22] as part of government engagement around natural capital which found that less than half the respondents were aware that collaborative platforms such as the NCC were available in their countries, suggesting that more needs to be done to engage with governments. Another noteworthy finding from the same study is that although government representatives agreed the private sector were key players, 81% could not identify a private sector champion for natural capital in their country, suggesting that more could be done to link private sector initiatives with those of governments. Complementary programmes do exist for governments, such as World Bank's WAVES, which are working to improve the development of country capacity by incorporating natural capital in national accounts, including natural resources such as forests, water, and minerals, as well as developing experimental accounts for ecosystems like mangroves [23]. Government buy-in for the NCP and natural capital accounting as part of national accounts will be a crucial driver for businesses to improve how they measure, value, and manage natural capital. Although business leaders are already using the NCP and aligned approaches, legislation and institutions will be required to 'level the playing field' and encourage the 'laggards' to proactively improve how they measure, and most importantly utilise, natural capital.

Bringing Change or Needing It?

Globally, significant advances have been made to raise the awareness of natural capital issues among business, governments, and individuals leading to the concept being discussed at major economic fora and generating a strong body of scientific literature and playing a central role in intergovernmental processes such as Intergovernmental Panel on Biodiversity and Ecosystem Services (IPBES). BES issues have featured for the first time in the top five risks of the global risk register published by the World Economic Forum [24]. In fact, three of the top five risks identified by WEF and ranked in terms of both likelihood and impact are categorised as environmental risks—although these have less significance to the current discussion as they focus on secondary impacts caused by extreme weather events, natural disasters, and a failure of climate change mitigation and adaptation. This trend is not entirely risk-centric either, with some lenders such as the International Finance Corporation incentivising the mainstreaming of natural capital into business models by offering preferential rates of interest [25].

Critics of the general commodification of nature, and—as a subset of the latter—the overly anthropocentric view taken by the natural capital approach, will be able to apply many of the same arguments to the application of the NCP. These concerns relate to attaching a monetary value to nature, whereby it could be traded and depleted at will where the short-term economic gains outweigh legal or moral obligations [26]. A recent article [27] goes further, arguing that this approach is a way to avoid hard choices and laws 'replacing political decisions with economic calculations'. Extrapolating this argument, Monbiot asserts that placing a pound or dollar sign in front of something is senseless as 'price represents an expectation of payment'; doing so is preparing nature to be sold and traded. Others argue that the natural capital framework—through the protocol—

'sets out an approach that reflects the full array of our inter-dependencies with nature, enabling them to be better reflected in decision-making' [28]. If appropriately applied, this could drive significant improvements on the current corporate and governmental decision-making trends, which rarely take biodiversity and ecosystem services into account.

In a similar vein, The NCC responds, in part, to concerns about the commodification of nature by presenting a more holistic definition of the term 'valuation'. In traditional accounting terms, the NCC clarifies, 'valuation is understood to mean monetization, but in environmental economics and this Protocol, valuation means more than just monetization. It includes qualitative, quantitative, and monetary approaches, or a combination of these' [18]. It goes on to present the framing of nature as 'natural capital' as 'a way of looking at the environment from an economic perspective. It views living and non-living natural resources as economic "stocks" from which a number of goods and service "flows" are derived'. The objective of such framing being a more detailed understanding of nature in economic terms, in a way that can be integrated into traditional decision-making processes which rely largely, some would argue disproportionately, on economic considerations.

Corporate reporting requirements, which in turn drive the formulation and application of internal company policy, including environmental policy, can play a role in broadening consideration of natural capital issues. These requirements are built around central principles which include 'materiality', which is both a general and a legal term broadly referring to issues that may have an impact on the issue(s) under evaluation, including the price of an asset or potentially its natural capital dependencies. One barrier here is the lack of a definition which includes reference to natural capital, which could prove challenging given that there is no common definition, with one assessment of experts reviewing eight, divergent, internationally accepted definitions of materiality [29].

'Placing natural capital and ecosystem services into a broader decision-making context is necessary to effect large-scale transformations in policies, practices, and investments' [1]. Challenges remain, however; natural capital remains overlooked, and the relationship between biodiversity and natural capital is poorly understood. This is significant as biodiversity underpins the functioning of ecosystems and plays a central role in managing and adapting to stress and damage to these systems from climate change, for example the recovery of coral reef ecosystems being an excellent case study. The need to better capture the multiple values of biodiversity and to integrate them into the NCP approach was laid out in a paper published by the Cambridge Conservation Initiative [28]. By focussing on the more obvious, business-relevant, relationships such as water, the resilience of these ecosystems—key to the sustainability of their resource flows—provided by biodiversity if at risk of being missed or undervalued.

A more specific consideration of the limitations of the NCP as a driver for change is that, as a tool for business, the scope and boundaries of the assessment are defined by the businesses conducting it, which will necessarily be chosen to benefit that specific business. Sensitive to uncertainty and indeed anything which may impact profits, as an obligation and legal commitment to their shareholders, this risks complicated issues being avoided and dependencies underplayed or not adequately addressed. This is not something the NCP itself was set up to address, however, and is an area where appropriate legislation and corporate policy must be formulated to incentivise more complete assessment, monitoring, and management of dependencies. The ultimate goal is to ensure that natural capital is fully integrated in national accounts, corporate policy, lending safeguard standards, and insurance policies. Full consideration of natural capital in such decision-making is likely to significantly change 'what' decisions are actually made [30]. While the argument for framing nature in economic terms so that it can be accounted for in current decision-making processes is sound, there is a paucity of voices calling for decision-making processes to themselves be changed as they appear to be fundamentally flawed—they do not currently take nature into account.

References

1. Guerry, A. D., S. Polasky, J. Lubchenco, R. Chaplin-Kramer, G. C. Daily, R. Griffin, M. Ruckelshaus, I. J. Bateman, A. Duraiappah, T. Elmqvist, M. W. Feldman, C. Folke, J. Hoekstra, P. M. Kareiva, B. L. Keeler, S. Li, E. McKenzie, Z. Ouyang, B. Reyers, T. H. Ricketts, J. Rockström, H. Tallis, and B. Vira.2015. Natural capital and ecosystem services informing decisions: From promise to practice. *Proceedings of the National Academy of Sciences* 112: 7348–7355. Available online at https://doi.org/10.1073/pnas.1503751112. Last Accessed 8 May 2018.
2. Science for Environment Policy. 2017. Taking stock: progress in natural capital accounting. In-depth Report 16 produced for the European Commission, DG Environment by the Science Communication Unit, UWE, Bristol. Available at: http://ec.europa.eu/environment/integration/research/newsalert/pdf/

natural_capital_accounting_taking_stock_IR16_en.pdf. Last Accessed 8 May 2018.

3. CBD. 2011. Aichi target 2. Avalabile online at: https://www.cbd.int/doc/strategic-plan/targets/T2-quick-guide-en.pdf. Last Accessed 15 May 2018.
4. Intergovernmental Platform on Biodiversity and Ecosystem Services (IPBES). 2012. Catalogue of assessments on biodiversity and ecosystem services. Available at catalog.ipbes.net. Accessed 21 May 2015.
5. Natural Capital Coalition. 2017. Natural capital protocol development. Available online at: https://naturalcapitalcoalition.org/protocol/development. Last Accessed 20 Apr 2018.
6. Hanson, C., J. Ranganathan, C. Iceland, and J. Finisdore. 2012. *The corporate ecosystem services review: Guidelines for identifying business risks and opportunities arising from ecosystem change. Version 2.0.* Washington, DC: World Resources Institute.
7. World Business Council for Sustainable Development (WBCSD). 2011. Guide to corporate ecosystem valuation. Available online at: https://www.wbcsd.org/contentwbc/download/573/6341. Last Accessed 9 May 2018.
8. Yergin, D. 1991. The prize, 559. New York: Simon & Schuster.
9. Schumacher, E. 1973. *Small is beautiful: A study of economics as if people mattered.* London: Vintage Publishing.
10. Missemer, A. 2018. Natural capital as an economic concept. *History and Contemporary Issues Ecological Economics* 143 (2018): 90–96.
11. Nunes, P.A.L.D., P. Kumar, and T. Dedeurwaerdere. 2014. *Handbook on the economics of ecosystem services and biodiversity.* Cheltenham: Edward Elgar Publishing.
12. Millennium Ecosystem Assessment (MEA). 2005. Ecosystems and human well-being: Synthesis. Washington, DC: Island Press. Mainstreaming the economics of nature: A synthesis of the approach, conclusions and recommendations of TEEB. Available at www.teebweb.org/publication/page/2/. Last Accessed 20 Apr 2018.
13. TEEB. 2010. Mainstreaming the economics of nature: A synthesis of the approach, conclusions and recommendations of TEEB, Available at www.teebweb.org/publication/page/2/. Accessed 20 Apr 2018.
14. Natural Capital Coalition. 2017. Natural capital protocol – History, vision and mission. Available online at: https://naturalcapital.org/who/history-vision-mission/. Last Accessed 30 Apr 2018.
15. Trucost. 2013. Natural capital at risk – The top 100 externalities of business. Available at: https://www.trucost.com/publication/natural-capital-risk-top-100-externalities-business/. Last Accessed 25 Apr 2018.
16. World Bank. 2014. Wealth accounting and the valuation of ecosystem services. WAVES Annual Report 2014, Available at www.wavespartnership.org/sites/waves/files/documents/WAVES_2014AR_REV_low_FINAL.pdf. Accessed 22 Apr 2018.

17. Natural Capital Finance Alliance (NCFA). 2012. Natural capital declaration. Available online at http://www.naturalcapitaldeclaration.org/. Last Accessed 8 May 2018.
18. Natural Capital Coalition. 2016. Natural capital protocol. Available online at: https://naturalcapitalcoalition.org/protocol/. Last Accessed 20 Apr 2018.
19. Natural Capital Coalition. 2017. Natural capital protocol finance sector guide. Available online at: https://naturalcapitalcoalition.org/protocol/sector-guides/finance/. Last Accessed 17 Apr 2018.
20. Natural Capital Coalition. 2017. Natural capital protocol application programme. Available online at: https://naturalcapitalcoalition.org/protocol/protocol-application-program/. Last Accessed 20 Apr 2018.
21. Natural Capital Coalition. 2017. Natural capital protocol yorkshire water case study. Available online at: https://naturalcapitalcoalition.org/natural-capital-protocol-case-study-for-yorkshire-water/. Last Accessed 20 Apr 2018.
22. Cambridge Institute for Sustainability Leadership (CISL). 2017. Snapshot of government engagement with natural capital approaches prepared to support the government dialogue on enabling natural capital approaches. World Forum on Natural Capital, Edinburgh available at: https://naturalcapitalcoalition.org/wp-content/uploads/2016/07/CCI-Natural-Capital-Paper-July-2016-low-res.pdf. Last Accessed 7 May 2018.
23. World Bank. 2017. World Bank WAVES programme. Available online at: https://www.wavespartnership.org/. Last Accessed 17 Apr 2018.
24. World Economic Forum (WEF). 2018. The global risks report, 13th edition. Available at http://wef.ch/risks2018. Last Accessed 8 May 2018.
25. International Finance Corporate (IFC). 2016. Press release: IFC lends €106 million to CEMEX environmental programs. Available online at https://ifcextapps.ifc.org/ifcext%5Cpressroom%5Cifcpressroom.nsf%5C0%5CD73AF0C19B1756AC85257FF7004E1816. Last Accessed 8 May 2018.
26. Bermejo Gomez de Segura, R. 2014. *Handbook for a sustainable economy*, 23. Dortrecht: Springer.
27. Monbiot, G. 2018. The UK government wants to put a price on nature – But that will destroy it. Guardian Online, Available at: https://www.theguardian.com/commentisfree/2018/may/15/price-natural-world-destruction-natural-capital
28. Cambridge Conservation Initiative (CCI). 2016. Biodiversity at the heart of accounting for natural capital: The key to credibility. Available at https://naturalcapitalcoalition.org/wp-content/uploads/2016/07/CCI-Natural-Capital-Paper-July-2016-low-res.pdf. Last Accessed 30 Apr 2018.
29. Corporate Reporting Dialogue. 2018. Statement of common principles of materiality of the corporate reporting dialogue. Available at: http://corporatereportingdialogue.com/wp-content/uploads/2016/03/Statement-of-Common-Principles-of-Materiality1.pdf. Last Accessed 7 May 2018.

30. Goldstein, J. H., G. Caldarone, T.K. Duarte, D. Ennaanay, N. Hannahs, G. Mendoza, and G.C. Daly. 2012. Integrating ecosystem-service tradeoffs into land-use decisions. *Proceedings of the National Academy of Sciences of the United States of America* 109 (19): 7565–7570. Available online at: https://doi.org/10.1073/pnas.1201040109. Last Accessed 1 May 2018.

CHAPTER 5

Debating Nature's Value: The Role of Monetary Valuation

Rob Tinch

Abstract Environmental valuation is grounded in expected utility theory. Coupled with the ecosystem services framework, it provides a quantitative model of how aspects of the natural world influence human well-being. Like any model, the important issue is not whether it is 'right' or 'true', but rather whether it is 'useful'. There are diverse possible uses, such as project appraisal, awareness raising, and supporting policy instruments. There are well-recognised problems, many applying also to market institutions and other forms of collective choice. The extent to which valuation is useful and acceptable will depend on environmental, economic, and social/political contexts. The key issue to debate is not whether monetary valuation is 'accurate', 'complete', or 'true', but rather 'under what conditions is monetary valuation useful?'

How should we view monetary valuation of environmental goods and services? The theory behind valuation is grounded in expected utility theory [10] and its more recent developments, which form an analytical framework used to explain people's decisions under uncertainty, based on the assumption that decisions stem from, and therefore reveal information about, individuals' preferences. In neoclassical economics, individual 'total

R. Tinch (✉)
Iodine sprl, Brussels, Belgium
e-mail: robert.tinch@iodine.be

V. Anderson (ed.), *Debating Nature's Value*,
https://doi.org/10.1007/978-3-319-99244-0_5

economic value' (TEV) represents all the ways that goods and services influence individual utility. This is revealed through the decisions or preferences of an individual, acting under a budget constraint, and expressed as their 'willingness to pay' (WTP). At a societal level, TEV represents the aggregate of these individual values, either as a simple sum or using weighting criteria, in particular to reflect income/wealth distributions and the diminishing marginal utility of income.

For a particular ecosystem or natural 'asset', therefore, TEV can be thought of as the sum of all the ways the ecosystem functions, ecosystem services and goods influence the utility of individual humans, as reflected by their WTP values, again either as a simple sum or following a weighting scheme (Fig. 5.1). Integrating TEV over time, using discounting to convert future values to present-day equivalents, gives the net present value of these flows. Assuming calculable risk about future flows, these values are often expressed as expected values, and cost-benefit analysis (CBA) compares the expected values of different courses of action. However, other treatments and decision rules may also be used, for example, to implement some degree of risk aversion in the calculations [11].

Even if nothing else, this provides a useful framework for thinking about ways that humans might value aspects of nature. Note in particular that, although the framework is grounded in individual preferences, it

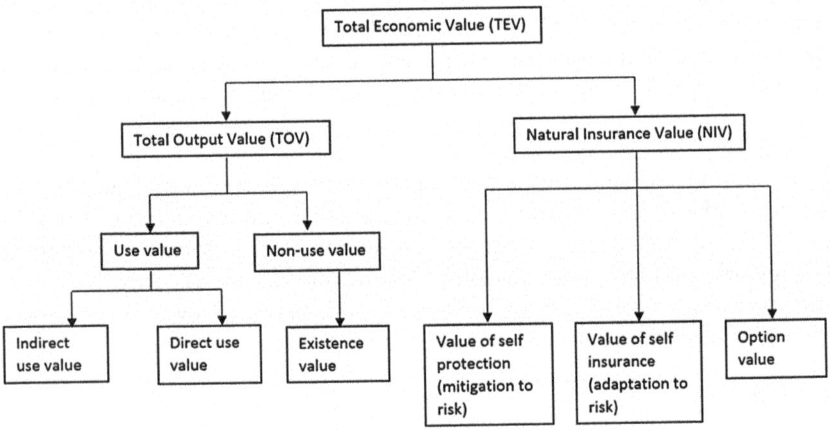

Fig. 5.1 The TEV framework (extended to include insurance values: see, e.g., Pascaul et al. [7])

nevertheless provides space both for non-selfish preferences (non-use values: existence, altruistic, bequest) and for uncertainty about future preferences and uses (option and insurance values). In a similar way, the ecosystem services framework (see e.g. [4]), often combined with the TEV framework, provides a useful checklist of ways in which natural systems provide benefits to humans. There is no claim that these values and benefits are an *exhaustive* representation of natural values; rather, the frameworks provide a minimum set of things to consider.

Thus, environmental valuation is one manifestation of a model showing how aspects of the natural world influence human well-being. Like any model, the important issue is not whether it is 'right' or 'true', but rather whether it is 'useful'. Thinking about it in terms of attempting to represent an underlying truth is not helpful, and makes for a rather easy straw man. It is much more interesting to consider whether or not valuation is useful as a decision support tool in different contexts. Here we should recognise that there are many different purposes and uses for valuation and CBA evidence. These include, for example, the following broad categories:

- Project appraisal, policy appraisal, and impact assessment
- Monitoring and review of decisions
- Demonstrating 'Value for Money', seeking funding
- Prioritisation of investments
- Planning and location decisions
- Pricing decisions: fees, payments, compensation for damages
- Understanding, communication, and advocacy

Each of these may call for different specific methods, and different requirements for accuracy and research expenditure, commensurate with the decision context and the spatial and temporal scale of application, ranging from localised, static appraisals to globalised, dynamic strategic assessments. Different applications in different social and political contexts may also evoke different ethical and practical objections.

There are many well-recognised problems, both theoretical and practical, with applications of the valuation model. Some of these are summarised in Table 5.1. While researchers are generally well aware of the limitations, and the results appropriately caveated (valuation and CBA guidance in particular always calls for sensitivity analysis, full reporting of

Table 5.1 Some problems arising in the valuation model

Assumption	*Problem?*	*Generalisation*	*Conclusion*
Individuals are the best judges of their own welfare	Demonstrably untrue in some cases (e.g. drug addiction) and doubtful in general (e.g. myopic decisions)	Democratic societies basically reflect this view and allow wide freedom of choice within a framework of rules to curb excesses	More a limit than a problem: recognise TEV focuses on individual preferences. Other moral decision rules may be considered
Individuals have the required information and cognitive ability to have stable, well-formed preferences that they express through decisions	Probably untrue, in particular for hypothetical decisions and unfamiliar goods and services, and preferences may be context-dependent and may vary over time	Poor information also affects other methods. Market institutions consistent with assumptions, with limits (advertising, trade descriptions, etc.)	Will reduce accuracy for some goods/services. Partial mitigation via information, time for reflection, deliberative method
Interpersonal comparability of utility	Not clear that any unit of benefit to one individual represents the same 'human welfare' as the same unit to another	A problem for any system (including voting systems) and not limited to monetary units	Practical option is to act 'as if' comparisons are reliable, using weighting to reflect priorities/ distributional goals
Values expressed are constrained by incomes/ability to pay	Derived estimates of social value assume that existing income distributions are desirable, or at least fair or that inequalities should be corrected through income policies	Policies to redistribute incomes via taxes and benefits mean that actual market distributions can be deemed at least in part a reflection of democratic decisions	WTP-based values for non-market goods and services are not necessarily valid measures of their social value; income weighting can help

(*continued*)

Table 5.1 (continued)

Assumption	*Problem?*	*Generalisation*	*Conclusion*
Smooth, continuous value functions	Non-linearities, threshold effects, and areas of highly inelastic demand/rapidly changing values	Severity depends on scale of application: small-scale, marginal assessments less likely to suffer than large-scale, major changes	Limits on use of valuation when dealing with critical natural capital or potentially catastrophic changes
Inevitable data gaps, in ecological/ scientific understanding, and/ or in the valuation evidence base	No valuation analysis or economic appraisal such as CBA can be considered complete and accurate	Applies to all methods, and CBA can include a wide range of values, sensitivity analysis, clear statements of gaps	CBA must be viewed as an aid to deliberation, not a way of providing 'the answer'
Optimism bias: tendency to underestimate future costs and overestimate benefits	CBA likely to be biased (both ways, including underestimating the costs and overestimating the benefits)	Not specific to economic valuation methods—more about physical outcomes and timings	Be aware of and make formal adjustment for optimism (or 'pessimism') bias

assumptions, weaknesses, omissions, etc.) this might not carry over to decision-makers using results.[1]

It is worth noting that many of the same criticisms apply (with varying force) to market institutions and indeed to other forms of collective choice. Markets using monetary currency as a unit of exchange and store of value are very powerful institutions for exchanging information about abilities and needs, and (nearly?) nobody would argue that we would be better off without them. But (not quite so nearly?) nobody believes that markets should be free of regulation or intervention—certainly not economists, who give considerable attention to market failures and possible remedies.

[1] Decision-makers misusing evidence to suit their ends is hardly unique: the question is whether monetary valuation evidence makes that easier or harder.

In any event, the use of market values to account for goods and services actually traded in markets (including ecosystem services such as food or timber production) is *relatively* uncontroversial—disagreements are mostly about rules and interventions (subsidies, taxes, redistribution, etc.), not the use of markets per se. But the estimation and uses of economic values for services such as clean air provision or biodiversity protection—or education, or health—can evoke very strong responses from different perspectives. In effect, the use of non-market valuation methods extends market thinking and tools to areas where property rights are not fully defined. This extension varies in degree, depending on the application, and can be very contentious, both on fundamental ethical principles and for practical reasons. For example, there is justifiable concern that valuation could support policies that are regressive, because it may appear more 'efficient' to cluster environmental 'bads' where people are poorer, because WTP is constrained by ability to pay.

The question of whether or not valuation 'helps' in any particular use or decision context seems to be the most important one. Valuation is certainly not essential: there are alternative ways of carrying out appraisal (e.g. Multi-Criteria Assessment (MCA), collective decision methods), and even environmental taxation could be implemented without necessarily using valuation to set the tax rates. But does it make these processes easier, more defensible, more transparent, more (cost-)effective? In particular, are arguments for recognising the importance of the natural world more convincing (for some decision-makers, in some contexts) if they are expressed in terms of monetary value?

Alongside that, we need to consider whether there are any unintended results, in particular over time. This is where concerns about 'crowding out' of non-market motives and values are important (see e.g. [8]). Similarly, is there a risk that expressing values in monetary terms provides a drive for those values to be 'captured' via market creation (i.e. defining property rights and bringing the environmental goods and services inside the 'productive boundary' of national accounts) and/or introduction of new environmental tax bases? And what would be the distributional impacts of that? Does use of valuation evidence create further demand for such evidence, locking decision processes into a particular approach (see, e.g., [6] in the UK water industry)?

These questions probably do not have single answers; rather, the extent to which valuation is useful will be dependent on environmental, economic, and social/political contexts. For example, there will always be bounds on the appropriate uses of values. Values change with quantities, so any particular point estimate is only going to be accurate at the margin, and any value for a non-marginal change in quantities is usually going to be an integral of a non-constant function. Thus the consequences of imprecise valuation depend on the elasticity of demand for an environmental service: risks are low where elasticity is low; where elasticity is high, rapidly changing values make the consequences of small quantity changes significant, so valuation and control by price are riskier [1]; for 'critical natural capital', elasticity is effectively infinite, marginal valuation is inappropriate, and the Precautionary Principle must apply [5].

So for me the key issue to debate is not whether monetary valuation is 'accurate', 'complete', or 'true', but rather 'under what conditions is monetary valuation useful?' Let me conclude with an example. The Costanza et al. [2] global valuation exercise is a popular punchbag, even amongst economists—the general reaction was 'these numbers are largely nonsense' but often also 'wish we'd thought of doing this'. And that is actually a pretty sensible reaction, because while nobody considers their ecosystem service valuation 'right',[2] the authors did largely achieve their aim of demonstrating that even a partial accounting of nature's gifts would show huge sums. But my main point is about their update [3]. The 1997 estimate for the global value of ecosystem services was c.$46 trillion/year in 2007 $US. The 2014 study updated the unit values and took account of land use/land cover change from 1997 to 2011. The new estimate is c.$125 trillion/year, but this does not imply that things have improved—it derives from an increase in unit values, not physical services. Rising values are not always a good sign: unit values will increase when a good or service becomes scarcer, and the net impact depends on elasticities. A revaluation of the 1997 services using 2011 values would be c.$145 trillion/year. So the repeated exercise showed the loss of eco-services from 1997 to 2011 due to land use change could be estimated at about $20 trillion/year. A meaningless number, or a powerful way of combining and communicating the combined impacts of a wide range of degradations that might otherwise be difficult to grasp and compare with other priorities?

[2] And they are certainly not the price or value of the natural world, nor do they claim to be.

Addendum: Values in Accounts

There is a lot of ongoing activity in the area of environmental accounting, so it is worth noting here that the purpose of accounting and the value principles being used are not the same as for welfare assessments. National accounting is what underpins GDP estimates, and omits most non-market activity, because 'balance has to be struck between the desire for the accounts to be as comprehensive as possible and the need to prevent flows used for the analysis of market behaviour and disequilibria from being swamped by non-monetary values' [9]. The national accounts do, however, include imputed values for several non-marketed services (including fixed capital consumption, healthcare, education, and the services of financial intermediaries) though not for others (notably services produced and consumed by households). Environmental and ecosystem accounting seeks to extend the boundary of the accounts to include ecosystems and their services.

National accounting uses exchange values ('prices'), not welfare values (TEV), and although GDP is often used as a measure of welfare 'there are several conventions in the SNA that argue against the welfare interpretation of the accounts' [9]. This explains why the environmental values used in accounts are different from those used in CBA. It also illustrates the risks of a tool being misused—and, indeed, of society becoming locked in to that misuse, with serious consequences (in the case of GDP, pathological focus on growing an indicator that is not even a measure of economic welfare).

It is this error that green accounting seeks, in part, to mitigate. But in this context, the interpretation of price changes in accounts is a particular concern. A higher figure in accounts might be thought 'better': more people benefit and/or there is an improved quantity or quality of service. But there are other possibilities: increased scarcity (overfishing, crop failure, etc.); decreased competition; loss of substitutes or increase in their prices; 'marketisation' (a self-fulfilling prophecy?). Sustainability requires maintaining welfare at lower throughput (flows), with non-declining wealth (stocks)—which is one reason why, if ecosystem accounts are to be used, it is vital that the physical accounts be considered alongside the monetary.

Acknowledgement This research has been supported by funding from the European Union's Horizon 2020 research and innovation programme under grant agreement No 678760 (ATLAS).

References

1. Baumol, W.J., and W.E. Oates. 1971. The use of standards and prices for protection of the environment. In *The economics of environment*, 53–65. London: Palgrave Macmillan.
2. Costanza, R., R. d'Arge, R. De Groot, S. Farber, M. Grasso, B. Hannon, et al. 1997. The value of the world's ecosystem services and natural capital. *Nature 387* (6630): 253.
3. Costanza, R., R. de Groot, P. Sutton, S. van der Ploeg, S.J. Anderson, I. Kubiszewski, S. Farber, and R.K. Turner. 2014. Changes in the global value of ecosystem services. *Global Environmental Change* 26 (2014): 152–158.
4. Daily, G.C. 1997. *Nature's services: Societal dependence on natural ecosystems*, 392pp. Washington: Island Press.
5. Farley, J. 2008. The role of prices in conserving critical natural capital. *Conservation Biology* 22 (6): 1399–1408.
6. Mathieu, L., R. Tinch, and A. Provins. 2016. Catchment management in England and Wales: The role of arguments for ecosystems and their services. *Biodiversity and Conservation* 27: 1–20.
7. Pascual, U., M. Termansen, K. Hedlund, L. Brussaard, J.H. Faber, S. Foudi, P. Lemanceau, and S.L. Jørgensen. 2015. On the value of soil biodiversity and ecosystem services. *Ecosystem Services* 15: 11–18.
8. Rode, J., E. Gomez-Baggethun, and T. Krause. 2015. Motivation crowding by economic incentives in conservation policy: A review of the empirical evidence. *Ecological Economics* 117: 270–282.
9. SNA. 2008. System of national accounts 2008, European Commission, International Monetary Fund, Organisation for Economic Co-operation and Development, United Nations, World Bank.
10. von Neumann, J., and O. Morgenstern. 1944. *Theory of games and economic behavior*. Princeton: Priceton University Press.
11. Wegner, G., and U. Pascual. 2011. Cost-benefit analysis in the context of ecosystem services for human well-being: A multidisciplinary critique. *Global Environmental Change* 21 (2): 492–504.

CHAPTER 6

Is the Concept of 'Natural Capital' Useful?

Tom H. Oliver

Abstract With growing political momentum and a keen enthusiasm to engage business and finance in environmental management, it is useful to review the merits and pitfalls of a 'natural capital' approach to ensure that a sensible trajectory is being pursued. Outlined in this chapter are some recommendations for a nuanced approach to the conceptualisation of 'natural capital' which takes into account the benefits and potential risks.

Natural Capital can be defined as the world's stock of natural assets which include geology, soil, air, water and all living things. It is from this Natural Capital that humans derive a wide range of services, often called ecosystem services, which make human life possible [5]. There is growing momentum behind a 'natural capital approach'; for example, in the UK reflected in the UK government's Natural Capital Committee, the Office for National Statistics' *Natural Capital Ecosystem Service Accounts* and the DEFRA 25-Year Environment Plan [2]. With growing political momentum and a keen enthusiasm to engage business and finance in environmental management, it is useful to review the merits and pitfalls of a natural capital approach to ensure that a sensible trajectory is being pursued. Following is a non-exhaustive list of considerations, along with

T. H. Oliver (✉)
School of Biological Sciences, University of Reading, Reading, UK
e-mail: t.oliver@reading.ac.uk

V. Anderson (ed.), *Debating Nature's Value*,
https://doi.org/10.1007/978-3-319-99244-0_6

recommendations for a nuanced approach to the conceptualisation of 'natural capital' which takes into account benefits and potential risks.

Benefits of the Conceptualisation of 'Natural Capital'

1. Defining and quantifying natural capital promotes awareness of our natural heritage and allows us to monitor whether it is being effectively protected. This also includes the ecosystem services that flow from natural capital, thereby maintaining key components underpinning human well-being. Note that the quantification of natural capital and its benefits does not necessarily require monetisation of the benefits provided (e.g. see UN System of Environmental-Economic Accounting; [4]).
2. The framing of the natural environment in terms of 'capital' puts it on an equal footing with other forms of capital (e.g. financial, manufactured, social, human capital), thereby highlighting its importance for the economy as well as our entire livelihoods. Using the language of economists is useful so that nature's values can be integrated into our macroeconomic frameworks.
3. Payment for ecosystem service schemes have been demonstrated to work in many cases (e.g. see [1]). These schemes rely on the quantification and valuation of natural capital and ecosystem services. For example, some water companies pay landowners to reduce nutrient inputs into watercourses. Both parties benefit financially from the arrangement, and there are also benefits for the environment, that is, a win-win situation facilitated by a natural capital approach.
4. The focus on natural capital and ecosystem services is a practical approach to environmental management. It is of course based on a utilitarian perspective (i.e. instrumental values) and ignores the 'intrinsic' value of nature. But can things really have an intrinsic value devoid of someone valuing them? Furthermore, species are just transitory phenomena (i.e. based on the 'background' extinction rate in the absence of humans each species only lasts between around 1 and 10 million years), so their value will be lost at some point anyway.

Disadvantages of the Conceptualisation of 'Natural Capital'

1. Some aspects of natural capital are harder to quantify and value than others. For example, valuing aspects of nature that provide cultural ecosystem services is notoriously difficult, whilst for provisioning services, such as timber, simple market prices can be used. Supporting services are very difficult to quantify, yet potentially underpin the basis of all natural capital benefits. As a consequence, these tend to get ignored in ecosystem accounting and consequent environmental management advice. As sociologist William Bruce Cameron once stated, *'not everything that can be counted counts, and not everything that counts can be counted'*. Such constraints are well recognised by organisations such as The Economics of Ecosystems and Biodiversity (TEEB) that have been seminal in developing valuation approaches for natural capital. However, despite clearly acknowledging our limited ability to accurately quantify aspects such as biodiversity value, in many cases they proceed to attempt monetary valuation of biodiversity anyhow, integrating it with other natural capital values to inform major environmental decisions (e.g. the Durban Metropolitan Open Space System in which a valuation of different aspects of natural capital preceded the development of an extensive international trade port). In other cases, less tangible aspects of natural capital such as biodiversity are simply ignored in monetary valuations.
2. Referring to nature as 'capital' implies it is equivalent to other forms of capital; yet it is fundamentally different. Some natural assets are not substitutable by other assets, that is, they are not *fungible*. Furthermore, with financial capital, we can invest it, or spend it to derive benefits, even borrowing capital to go into debt. But we cannot do the same for natural capital. Treatment of natural assets in identical ways to other capital would be catastrophic for the environment.
3. Beyond a narrow subset of economists and policymakers, people do not engage with technical terminology such as 'natural capital' and 'ecosystem services'. Thus, using such phraseology is not a good way to convince people to adopt pro-environmental behaviours or justify environmental decisions to citizens.

4. Natural capital lends itself to financialisation of nature. In some cases, there may be benefits of economic incentives (e.g. rewarding farmers to provide public goods or providing insurance products to protect farmers and associated industries from pro-environmental practices that are more risky financially). However, there is danger in financialisation of nature in large open markets. For example, natural capital offsetting schemes, whereby damage to nature in one location is offset by habitat restoration elsewhere could in theory be translated into large, even transnational, markets. However, such global markets are subject to huge volatility, especially when financial innovations extend beyond simple useful products (e.g. providing capital and insurance) to highly complex structures with no clearly defined social benefits (e.g. collateralised debt obligations responsible for the sub-prime mortgage crisis and 2008 financial crash). The irreversibility of many natural assets makes unstable markets highly dangerous. Despite this, many entrepreneurs see potential profit is such marketisation of nature, and they may lobby strongly for its development.
5. National natural capital approaches need consideration in an international context. A limited focus on accounting and development of national markets may lead to offshoring of social and environmental impacts (e.g. protecting cultural ecosystem services locally through extensive habitat restoration whilst importing unsustainably produced food and energy from overseas). If markets are globalised, consideration needs to be taken of the inequity in spending power between countries. Not all ecosystem services are equally essential for humans; for example, cultural ecosystem services are arguably less immediately essential than food provision (cf. Maslow's hierarchy of needs). In completely free markets there is a danger that wealthier countries might satisfy non-essential needs at the expense of essential needs in poorer countries.

A Nuanced Approach to the Conceptualisation of 'Natural Capital'

Taking into account the mentioned advantages and disadvantages, rather than solely rejecting or wholeheartedly embracing the concept of natural capital, one may consider its useful application in certain contexts. Some points for consideration of such a nuanced position are outlined as follows:

1. The effectiveness of a given message depends on the audience. There is no need to use the same message for all audiences. For policymaking audiences, economic costs and the benefits of alternative options are most useful, whilst for engendering behaviour change in the general public, positive emotional messages may be most effective (e.g. see [3]).
2. A natural capital approach can be adopted to a limited degree without wholesale marketisation of nature. So, for example, it might be used for national accounting, and valuation might be used for local closed markets (e.g. auctions within individual water catchments, where landowners bid to devote land to provide nature-based water purification or flood protection in return for monetary payments). However, large markets for natural capital with higher risks need to be avoided and caution may be needed that vested interests may try to push for a broader application.
3. If a natural capital approach is adopted, goods that cannot be adequately quantified and valued in monetary terms must not be ignored; they need to have an equal footing in decision-making processes. For example, they may act as constraints (e.g. 'no net loss of biodiversity') or other non-monetary valuation approaches may be used, such as deliberative participatory approaches or using multicriteria decision analysis.
4. Adoption of market-based solutions under a natural capital approach should not cancel out or overshadow other approaches to environmental protection (such as new environmental regulation and improved policing of existing regulation). The academic community could be more active in voicing concerns that if demands on the environment continually ratchet upwards, then no amount of valuation and optimisation of ecosystem service delivery can fully solve environmental problems. Therefore, spatial information approaches

to the management of natural capital (cf. [2]) are only half of the solution, and more systemic approaches to reducing environmental impacts (e.g. energy efficiency and waste reduction) are equally essential.

5. More attention should be given to international context and the potential offshoring of environmental and social problems through a natural capital approach. These approaches have been arguably lacking in recent policy approaches (e.g. [2]).

References

1. DEFRA. 2013. Payments for ecosystem services: A best practice guide. Annex-Case studies. https://www.gov.uk/government/uploads/system/uploads/attachment_data/file/200901/pb13932a-pes-bestpractice-annexa-20130522.pdf
2. DEFRA. 2018. A green future: Our 25 year plan to improve the environment. https://assets.publishing.service.gov.uk/government/uploads/system/uploads/attachment_data/file/693158/25-year-environment-plan.pdf
3. Futerra. 2015. Branding biodiversity – The new nature message. Futerra sustainability communications. http://www.wearefuterra.com/wp-content/uploads/2015/10/Branding_Biodiversity.pdf. Accessed 27 Feb 17.
4. SEEA. 2017. The system of environmental-economic accounting (SEEA). Experimental ecosystem accounting briefing paper. https://unstats.un.org/unsd/envaccounting/workshops/int_seminar/note.pdf. Accessed 27 Jan 17.
5. World Forum on Natural Capital. 2015. http://naturalcapitalforum.com/about/

CHAPTER 7

How Should We Value Nature?

Sandra Bell

Abstract Nature is in crisis and urgently needs better protection. Our own well-being depends upon thriving nature. Economists say that nature will be better protected if we can assess it in monetary terms. This seems appealing—nature has been undervalued for too long. But there are risks and flaws in the Natural Capital approach—the way we experience nature or the simple right of nature to exist cannot be measured in this way. And putting a monetary value on nature's services is risky—would we accept the loss of most bee species as long as the species that do most crop pollination are saved? Natural Capital accounting may be a useful tool in some limited circumstances but it is no substitute for strong regulation and policy reform.

As non-governmental organisations (NGOs) we are facing increasing challenges to convince politicians at all levels to make the right decisions for nature. Governments are beginning to wake up to the financial value of nature but there are risks and flaws in the approach being taken. We need to ensure that the true value of nature to all of us is recognised and taken into account in decision-making.

S. Bell (✉)
Friends of the Earth, London, UK
e-mail: Sandra.Bell@foe.co.uk

V. Anderson (ed.), *Debating Nature's Value*,
https://doi.org/10.1007/978-3-319-99244-0_7

How Economists Value Nature

The economists who are now taking an interest in nature say that we have failed to reverse biodiversity decline because we have failed to recognise the values of nature. Pavan Sukhdev (The Economics of Ecosystem and Biodiversity)[1] stated that '[t]he crisis of biodiversity loss can only begin to be addressed in earnest if the values of biodiversity and ecosystem services are fully recognised and represented in decision making'.

To some extent, Friends of the Earth agrees with this analysis—nature underpins our economy yet gets scant recognition for all the services it provides us with for free. But we have serious concerns about how this has been taken on board. For example, the Natural Capital Committee talks about fully integrating the environment into the economy and its Chair, Professor Dieter Helm, suggests the prioritisation of protection of nature by 'focusing on those areas with the highest economic benefits'.[2]

There are many reasons for which we should not value nature only in economic terms, not least that as humans we have a connection with nature that cannot be described in a spreadsheet. We cannot value nature for its intrinsic value if we have to always put a price on it.

It is perhaps telling to look at the list of environmental services that are in the 'not included' list in the UK's natural capital monetary estimates from the Office of National Statistics (ONS)[3]: among these are 'Value placed on nature simply existing', 'Heritage and aesthetic interactions', 'wild animals and 'Symbolism of certain plants and animals'.

[1] TEEB (2010), The Economics of Ecosystems and Biodiversity: Mainstreaming the Economics of Nature: A Synthesis of the Approach, Conclusions and Recommendations of TEEB. http://www.teebweb.org/publication/mainstreaming-the-economics-of-nature-a-synthesis-of-the-approach-conclusions-and-recommendations-of-teeb/

[2] Natural Capital Committee (2015) Third report to the Economic Affairs Committee. https://www.gov.uk/government/collections/natural-capital-committee-documents

[3] Office of National Statistics (2016) UK natural capital: monetary estimates https://www.ons.gov.uk/economy/environmentalaccounts/bulletins/uknaturalcapital/monetaryestimates2016

The Value of Bees

Even from a utilitarian perspective of what nature does for us there is a high risk in using the short-term approach of economics and applying it to nature and natural systems. For example, research by the University of Reading in the UK[4] shows that we have the ability to identify a handful of bee species that currently contribute the highest economic returns in terms of providing pollination services to crops. But in a changing climate we may need different bee species to pollinate our crops in the future. So a decision to prioritise conservation of the most useful species now and ignore the rest could be disastrous for our future food security. As Professor Simon Potts said, '[p]utting a cash value on ecosystem services is helpful to highlight to politicians and farmers just how important nature is to the bottom line, … [b]ut thinking purely about today's profits is pointless if it comes at the expense of the future sustainability of our countryside and our food supply'.

How do we value the majority of bee species that may have a smaller role to play in crop pollination or do not pollinate our crops at all? These species play a vital role in pollinating the plants that other wildlife depends upon and our own experience of nature is closely linked to pollinators, whether it is hearing buzzing bees or enjoying the flowers they pollinate. The conclusion of a study in the role of different pollinator species in Europe and North America[5] concurred that there is a risk in putting too much focus on an ecosystem service approach when it comes to wild pollinator conservation. The authors concluded that '[m]oral arguments remain pivotal to supporting conservation of the larger portion of biodiversity including threatened species that currently contribute little to ecosystem service delivery. Such arguments are powerful and define many human actions, from taking care of the elderly to preserving historical buildings or art. Ecologists and conservationists need to make these distinctions clear if we expect policy makers or land owners to defend species with no clearly defined economic value to humans'.

[4] University of Reading (2015) Britain needs 'super-sub bees' to maintain food supply https://www.reading.ac.uk/news-and-events/releases/PR635710.aspx

[5] Kleijn et al. (2015) Delivery of crop pollination services is an insufficient argument for wild pollinator conservation, Nature Communications https://www.nature.com/articles/ncomms8414

Friends of the Earth does not reject every connection of nature to the economy. But we do reject the notion that decisions about protecting nature are made on the basis of its contribution to economic growth alone, and we are concerned about how natural capital is currently being measured.

How People Value Nature

When we ask people why they care about nature they have an emotional response—they never mention monetary value. This is just a small selection of comments from Friends of the Earth supporters:

> **"Saw puffins for the first time by the score, also fulmars, kittiwakes, razorbills, greater black backed gulls. The place is magic. I wish I could live there".**
>
> **"Being in nature restores my sanity half the time! Even if it's just in the back garden. The smells, sights sounds and feelings of the nature just to be found there is wonderful".**
>
> **"Standing knee deep in the fast flowing streams in the black of a warm July night, as the sea trout run the river – it's a great experience".**

Everyone should have access to high-quality, thriving nature. But too often it is marginalised, poor and/or ethnic minority communities with the least power and money who are denied the ability to enjoy and experience nature, and the proven benefits that nature brings us. People living in areas with high deprivation are more likely to have less access to green spaces.

The natural capital approach risks that the value of accessing nature is measured, not for its contribution to our well-being but for its recreational value, assessed in monetary terms (e.g. money spent on travel and entry fees as in the ONS assessment).

Nature Should Not Be Traded

Friends of the Earth is also concerned that assigning monetary values or metrics to nature allows it to be traded—in schemes such as biodiversity offsetting—which puts forward a flawed notion that nature, in all its complexity, is something that can be measured, exchanged and replaced in another location.[6]

Offsetting is now being introduced in the UK. A housing proposal at Lodge Hill in Kent could destroy a legally protected wildlife site, one of the last remaining strongholds for nightingales in the UK. The council considered that the nightingales could simply move to newly created habitat some distance away and approved the application (the application was subesequently withdrawn but the future of this site still hangs in the balance).

There are many risks and flaws with an offsetting approach, including that the value to people of having access to nature on their doorsteps is not considered when habitat is shifted elsewhere. There is little or no evidence from existing schemes that offsetting delivers gains for biodiversity—nature cannot simply be shifted around from one location to another.

The natural capital approach is now embedded in the UK Government's much anticipated 25-Year Plan for the Environment. On the positive side the advice of the Natural Capital Committee probably helped to raise the ambition of the plan.[7] But there are also concerns about the kind of tools that are being proposed and in particular the concept of Net Environmental Gain. Net gain is fraught with problems and risks contributing to environmental decline instead of gain. For example, will developers be allowed to harm habitat because they meet energy standards which should be met anyway?

A New Debate About Nature Is Needed

If we want to ensure that politicians value nature in their decision-making beyond a price tag, we need a new public debate to frame public understanding of the importance and relevance of nature. We need many more people standing up for nature including health professionals, teachers and parents concerned about their children's health and (lack of) contact with nature.

Whilst we want responsible businesses to understand the value of nature we should not always have to justify nature protection as making good business sense, or being 'good for the economy'. Nature should be pro-

[6] https://www.foeeurope.org/nature-not-for-sale

[7] https://www.gov.uk/government/publications/natural-capital-committee-advice-on-governments-25-year-environment-plan

tected for its intrinsic value—and also for the wider immeasurable benefits it brings to people—our sense of wonder and our health and well-being in being connected to nature.

Nature is irreplaceable. So the idea that habitats and species are tradable and that damage in one place can simply be replaced in another place, as in biodiversity offsetting proposals, is a breach of people's and communities' right to local nature in their lives and on their doorsteps. Trading biodiversity for other, easier-to-measure, environmental benefits would be a disaster for nature.

Assessing the economic value of some aspects of nature is a useful tool that helps us understand some of the benefits that nature brings us but it can never tell us the true and total value of nature, so it should not become central to the way in which decisions are made about nature protection and restoration. It would be wrong to prioritise nature protection on the basis of greatest economic returns.

Real Solutions

The ultimate way to protect nature is to address the core issues and drivers that cause it to be damaged or lost. This must include redirecting public money to support farming that boosts nature instead of propping up damaging intensive farming practices, and the proper implementation of existing international obligations and European Union (EU) laws, including the Birds and Habitats Directives. Strong national land use planning policies and good governance at the local level will also be essential—supported by independent nature watchdogs and local ecological expertise. In the UK post Brexit it will be essential that we not only maintain environmental regulations and principles from the EU but improve their implementation and enforcement.

More information:

https://friendsoftheearth.uk/nature
https://www.foeeurope.org/nature-is-our-right
http://www.natures-keepers.org/

CHAPTER 8

Natural Capital: The Risks of Losing Sight of Nature

Rebecca Clark

Abstract This chapter looks at the prominence of nature in applications of natural capital, focusing on the steps that the UK's Natural Capital Committee has suggested for developing a long-term plan for investment in the environment. The chapter finds that there is a risk of losing sight of nature and focusing on delivering environmental goods and services to people.

Introduction

Natural capital provides a lens for viewing the environment: our reliance on it, the impacts we have on it and the case for maintaining and improving it. In the UK, the work of the Natural Capital Committee[1] (NCC) has raised the profile of the concept. The Committee defines natural capital as

[1] An independent advisory body to the UK government established in 2012.

R. Clark (✉)
Natural England, Norwich, UK
e-mail: Rebecca.Clark@naturalengland.org.uk

V. Anderson (ed.), *Debating Nature's Value*,
https://doi.org/10.1007/978-3-319-99244-0_8

> *[t]he elements of nature that directly or indirectly produce value to people, including ecosystems, species, freshwater, land, minerals, the air and oceans, as well as natural processes and functions* [4]

This definition combines two components: (a) provision of goods and services that provide benefits to people[2] and (b) processes and functions of nature (or ecosystems). The inherent assumption is that a healthy natural environment is required for long-term sustainable delivery of goods and services. However, in applications of natural capital, is a healthy natural environment still prominent? Or is there a risk of losing sight of nature and focusing on delivering environmental goods and services?

An important feature of the NCC's approach to natural capital is that it considers stocks of natural capital ('assets')[3] as well as the flow of goods and services[4] that they deliver. Degradation of natural capital assets has potential to impact on delivery of goods and services. This could affect economic activity (e.g. through increased flooding), direct benefits to people (e.g. places for recreation) and people's health (e.g. through poor air quality). Degradation may also have impacts that are not yet understood and affect benefits to people in future that have not yet been identified (e.g. genetic material from wild plants providing resources for crop breeding in future). Because of this, the NCC [3] advises that degradation of natural capital is best avoided.

Through international agreements (e.g. to the Aichi Biodiversity Targets) and national policy and legislation, society has demonstrated that conservation of nature is a priority. This is for ethical and other reasons. However, impacts on the environment are easily overlooked in decision-making which can affect the outcomes and effects. This is because many of the goods and services provided by the environment are not bought and sold (e.g. a fine view, purification of water). Monetary values can be

[2] These goods and services are used directly by people (such as places for recreation) or as inputs to production processes, either in households (such as preparation of food) or businesses (such as production of timber). When the outputs are used or consumed, they provide benefits to people.

[3] The classes of assets used by the NCC [4] are species, ecological communities, soils, land, freshwater, coasts, oceans, atmosphere, minerals and subsoil assets.

[4] The categories of goods and services used by the NCC [3] are food, fibre, energy, clean water, clean air, recreation, aesthetics, wildlife, protection from hazards and equable climate.

estimated for some of these (e.g. recreational visits, carbon sequestration) using various techniques. If impacts on goods and services that can be valued are considered in decision-making, effects on the environment are partially taken into consideration. However, the impact on nature is not adequately reflected because it can only be partially assessed in economic terms.

Decision-making that is focused on delivering environmental goods and services will not necessarily deliver a healthy natural environment (though it may do so in specific circumstances). Take the example of a newly restored wetland that is managed with an objective of delivering a recreational resource for people, for example, by providing access for water sports and year-round use of an extensive network of footpaths. Such management would deliver some environmental benefits (by retaining the wetland). However, it could also impact negatively on nature through disturbance to birds on the water and to nesting birds on the ground amongst other things. Consequently, decision-making needs to consider the effects on nature as well as impacts on environmental goods and services.

The following five sections consider the prominence of nature in the steps for developing a natural capital investment plan that were set out by the NCC [5]. Such a plan could be developed by the government or a business.

Steps for Developing a Long-Term Investment Plan for Natural Capital

Measuring and Monitoring Changes in Natural Capital

Monitoring is employed to measure change in natural capital over time. The prominence of nature is dependent on the purpose of the monitoring and the metrics that are used. The NCC [5] recommends that for management and investment purposes monitoring is focused on delivery of environmental goods and services. Addition of metrics on ecosystem quality, quantity, location, functioning and thresholds will help ensure that nature is also monitored.

Integrating Natural Capital into National and Corporate Accounts

The purpose of integrating natural capital into national and corporate accounts is to encourage sustainable management of natural capital in the long term:

- The ecosystem accounts (that are being developed by the Office of National Statistics) that accompany the UK's national accounts[5] will provide long-term monitoring of the nation's natural capital assets. They will provide insight into the impacts of economic activity on natural capital and reliance of that activity on the environment.
- The framework for corporate natural capital accounts (CNCA) developed for the NCC [5] aims to help businesses monitor long-term maintenance of the natural capital assets that they are responsible for. Amongst other things, it can also help reveal dependencies on natural capital that could pose risks to the business and potential sources of revenue from environmental goods and services.

In the national ecosystem accounts and CNCA a component of the account is focused on the quantity, quality and location of ecosystem(s) and their functioning. Both also include components with information on the quantity of goods and services delivered by the environment and the value of the goods and services. Nature is prominent in the ecological indicators in both accounts. However, the temptation to focus on monetary values means that these are easily overlooked. The focus is then on the value of environmental goods and services.

CNCA differs from the national accounts in that it also includes costs of managing the environment. Currently, the national accounts do not include such estimates of the costs, though it has been argued that they should (e.g. [2]). The inclusion of costs is intended to help secure maintenance or investment in natural capital in future [5]. It could therefore have a significant effect. The long-term environmental objective used in the account (which is specified by the government or business) determines

[5] These measure economic growth in terms of gross domestic product, amongst other things.

the costs that are included.[6] If natural capital is understood to include nature as well as delivery of goods and services, the objective will aim to support both. This is in contrast to an objective that is focused on delivery of goods and services (which will only partially support nature).

Including Impacts on Natural Capital in Cost-Benefit Analysis[7]

Decision-making by businesses and by the government is often informed by cost-benefit analysis (CBA). This helps identify options that provide best use of available resources and deliver value for money. If impacts on the environment are not incorporated in this analysis, they may be assumed to have no value. Decision-makers then consider environmental impacts separately to the CBA findings and there are risks that low weight is put on the impacts or they are overlooked.

To help ensure that environmental impacts are considered, the NCC [5] recommends that impacts on environmental goods and services be valued if possible and included in CBA. Environmental impacts can only be partially incorporated into CBA because only some goods and services can be valued (as discussed in the introduction). However, including those that can be valued can alter the findings from the analysis (e.g. inclusion of recreational benefits of woodland in economic analysis reveals that planting near cities delivers higher benefits compared with remote locations [1]).

Impacts on nature that do not directly affect goods and services cannot easily be valued. Because no value is put on these, they are not included in CBA. Consequently, for nature to remain prominent, impacts on nature need to be considered in decision-making and given equal weight as CBA findings (even if the CBA includes impacts on environmental goods and services).

[6]The NCC [5] suggests that private organisations should be responsible for maintaining the natural capital assets that they own in perpetuity with no loss in the assets. This builds on the concept that the current generation holds the management of natural capital in trust for future generations.

[7]The issues discussed in this section also apply to the appraisal of costs and benefits of specific options for investing in natural capital, which is a separate step identified by the NCC [5].

The issues identified also arise in another of the steps for developing an investment plan: appraising costs and benefits of specific options for investing in natural capital.

Defining Outcomes and Targets

Specifying outcomes and targets for natural capital will help inform development of an effective plan for long-term delivery. At a national level, the NCC [5] suggests that the targets should be at least those required to deliver commitments in international and domestic policy; more is needed to secure the Government's commitments to improve the environment. If organisations individually specify the contributions that they plan to make, this will help secure delivery at a national scale and identify shortfalls.

The prominence of nature in outcomes and targets depends on the long-term environmental objectives that are being pursued by the government or a business. Resilient ecosystems will not necessarily be delivered by outcomes and targets that are focused on delivery of goods and services. If targets are set based on scientific criteria, policy and legislative commitments, they may include nature. Targets that are based on CBA will focus on goods and services that can be valued.

Developing a Risk Register

The NCC [3] developed a risk register that identifies environmental goods and services that are at greatest risk (due to the ways they are being used or managed) and the assets that are most important in delivering them. Risk was assessed by experts on the basis of delivery of relevant targets (e.g. set out in policy) and known thresholds. This approach was focused on goods and services. Risks to nature could also be assessed, for example, if the approach included risks to the quality, quantity and location of ecosystems. The prominence of nature is also influenced by the targets that are used to assess risk (it will be lower if the targets are focused on goods and services).

Conclusions

The prominence of nature in applications of natural capital is considered here in terms of the stages for developing a long-term investment plan developed by the first UK NCC. Environmental goods and services that provide benefits to people feature heavily in all of the stages. Nature is prominent in a few of them and could be incorporated into some others. The emphasis placed on nature is strongly affected by the long-term environmental objectives of the government or business that is undertaking the work. There is a risk of losing site of nature and focusing on goods and services, of not being able to see the trees for the wood.

References

1. Bateman, I., B. Day, M. Agarwala, P. Bacon, T. Bad̆ura, A. Binner, A. De-Gol, B. Ditchburn, S. Dugdale, B. Emmett, S. Ferrini, C. Carlo Fezzi, A. Harwood, J. Hillier, K. Hiscock, M. Hulme, B. Jackson, A. Lovett, E. Mackie, R. Matthews, A. Sen, G. Siriwardena, P. Smith, P. Snowdon, G. Sünnenberg, S. Vetter, and S. Vinjili. 2014. *UK national ecosystem assessment follow-on. Work package report 3: Economic value of ecosystem services*. UNEP-WCMC, LWEC, UK.
2. Mayer, C. 2013. *Unnatural capital accounting*. Paper for the Natural Capital Committee.
3. NCC. 2014. *The state of natural capital: Restoring our natural assets*. Second report to the Economic Affairs Committee.
4. NCC. 2014. *Towards a framework for defining and measuring changes in natural capital*. Working paper 1.
5. NCC. 2015. *The state of natural capital: Protecting and improving natural capital for prosperity and wellbeing*. Third report to the Economic Affairs Committee.

CHAPTER 9

Some Difficulties of Measurement

Victor Anderson

Abstract Even if one wanted to measure the value of 'natural capital' in monetary terms, could it be done? Can it actually be measured without such a wide range of values being given that doing so is effectively useless for any decision-making processes? This chapter explores a series of difficulties about the measurement of 'natural capital'. All these derive principally from the problem that 'natural capital' is a metaphor which doesn't quite fit. The natural world and the world of capital are different in important ways.

The main focus of this book is on the many different implications of attempting to measure the value of 'natural capital' in monetary terms. However, in this chapter I will explore a different question: even if one wanted to do it, can it be done? Can the monetary value of "natural capital" actually be measured without such a wide range of values being given that doing so is effectively useless for any decision-making processes?

There are in fact a whole series of difficulties about attempting to carry out these measurements. None of them provides an argument against the use of the term 'natural capital' in a general way, for example as a form of metaphor or rhetoric, or as a slogan or a conceptual bandwagon to be

V. Anderson (✉)
Global Sustainability Institute, Anglia Ruskin University, Cambridge, UK
e-mail: Victor.anderson@anglia.ac.uk

V. Anderson (ed.), *Debating Nature's Value*,
https://doi.org/10.1007/978-3-319-99244-0_9

climbed aboard temporarily as a source of political respectability and funding, but they do cause problems for anyone given the task of actually following that through into trying to work out the figures.

Many of the difficulties arise from the 'natural capital' metaphor itself. It does not quite fit. In some ways the natural world can be seen as a source of a flow of benefits, and in that sense similar to capital. At the same time, there are other respects in which it is crucially different. Discussion and analysis based on assuming away those differences are almost certain to be unrealistic. Consideration of the problems of measuring 'natural capital' can usefully start from considering the ways in which it differs from capital in the ordinary economic senses of the word (machinery, buildings, money). There are two particularly important differences which are focused on the ideas of 'divisibility' and 'substitutability'.

1. Capital in the sense of money is said to be 'infinitely divisible'. There may be practical problems about getting to fractions of pennies or cents, but in principle money may be of any quantity. Capital in the sense of machinery or buildings is not divisible to the same extent, but it does make sense to talk of particular separate buildings and particular separate machines. However 'natural capital' is not like that. There are no 'pieces' of natural capital which can each be valued separately. The different parts of the natural world are inescapably interdependent, and that understanding is the basis of the whole idea of ecology. It makes no sense, therefore, to abstract a particular species from its ecosystem and then ask how much it is worth, because its value depends very largely on the role it plays in that ecosystem.
2. Capital in the sense of money is interchangeable. Given the existence of international exchange rates, it is interchangeable across the globe. The natural world is very different: it consists of particular places, each with their own specific character. They are not easily interchangeable. Differences are qualitative as well as quantitative. Assigning a monetary value to a 'piece' of 'natural capital' suppresses its qualitative aspect, its uniqueness and specificity.

Three further problems arise from biases in the methods often used for monetary valuation, without being inherent in the idea of monetary valuation itself. In some cases, methodologies are used to correct for these biases, but they do nevertheless create a source of difficulty for carrying out measurements of value because these biases can be handled in different ways, with different outcomes.

3. Theoretical valuation is biased against poorer people when it is based on 'willingness to pay', because poorer people do not normally imagine being able to pay out sums as large as richer people do. In that sense, the theoretical market of monetary valuation tends to reproduce the bias of the real market.
4. Valuation is often operated on a 'net present value' (NPV) basis—which involves adding up the costs and benefits at different points in time, applying a discount rate to costs and benefits which occur in the future. The main justification used for this is that people have a 'time preference': we would rather have a pound right now than even a 100% certainty of receiving a pound in five years' time. The argument is that this should be reflected in economic calculations, treating future costs and benefits as of less value than costs and benefits in the immediate future. The rate at which future costs and benefits are reduced by comparison with those same costs and benefits in the present is known as the 'discount rate'. The use of anything larger than tiny discount rates more or less guarantees that costs and benefits a generation or more into the future will count for virtually nothing. Since sustainability is all about what can be sustained into future generations, the use of discount rates for valuation is not an approach which gives any priority at all to sustainability, and in practice it can lead to conclusions which are extremely unsustainable. This creates a form of valuation with a built-in bias against the interests of members of future generations. As discussion about the Stern Review on climate change economics has shown, the choice of discount rate is crucial for the policy conclusions drawn.[1]
5. A bias in favour of the owner of a 'piece of natural capital' can slip in if there is a lack of effort to avoid confusion between the value to the owner and the value to society as a whole. A firm can claim to have

[1] See Lawrence H. Goulder and Roberton C. Williams III: 'The Choice of Discount Rate for Climate Change Policy Evaluation' (Resources for the Future 2012). They say (page 1): 'Nordhaus indicated that the differences between the Stern-endorsed and Nordhaus-supported discount rate accounted for all of the difference between the more aggressive climate policy endorsed by Stern and the considerably more modest effort supported by Nordhaus'.

gone beyond the scope of traditional accounting of profit and loss, assets and liabilities, to also consider its 'natural capital'—and it might hope to claim some 'corporate social responsibility' credibility for doing so. This might be encouraged by manuals providing guidance to companies as to how to go about valuing their 'natural capital'. However, what is a natural capital asset to a firm (e.g. an oil well) might be a cost to society as a whole (e.g. in the form of carbon emissions). Where natural capital valuation is used, there will still be a need to make this important distinction between 'private' and social forms of costs and benefits. It will often follow from this that there is a strong case for government intervention or a transfer of ownership, and natural capital valuation will not have removed the need for these changes to be considered.

Finally in this catalogue of difficulties, there are two problems which derive from differences between the actual market and what can be described as the 'theoretical market' of natural capital valuation.

6. Pieces of capital (in the ordinary senses of the word) have particular prices in the market. However, valuation studies attempting to put money values on a 'piece' of natural capital often reveal a very wide range of different values placed by different groups of people. There is no shared sense of a 'value' or resolution of conflicts about what value should be assigned unless an actual market is established. Any real market 'value'—or more accurately, price—for something depends largely on the relative bargaining and purchasing power of different groups within that market. It is impossible to specify a money value without that process having taken place, or at least been imagined and allowed for. Valuations depend on some set of rules being established for reconciling the different values assigned by different groups. There might, for example, be procedures involving majority voting or taking averages. Varying these rules will of course tend to vary the values arrived at.
7. Values assigned through valuation studies depend a great deal on consumer preferences, and these are in constant flux. Things which appear important and valuable to consumers can differ a great deal from one year to the next. Things which are dramatic, fashionable, receive publicity or are easily understood may appear more valuable than things which are obscure. For example, worms vital to the soil's capacity for producing food may be assigned a tiny value in

comparison with some well-known charismatic mammal. This does not seem like a secure basis for decision-making and policymaking, which should be based at least in part on facts about the world, or at least as close to them as current science is able to get.

The most obvious conclusion from this list of basic difficulties is the need to be extremely cautious about the use of monetary valuation and the application of the concept of 'natural capital'. It has its uses as a metaphor and piece of rhetoric, and for some audiences is a means of communication and persuasion, but as soon as it is taken literally and someone is given the task of actually calculating the figures, it falls into a whole series of problems, principally based on the fact that the metaphor does not quite fit.

CHAPTER 10

Who Should Value Nature?

Dario Kenner

Abstract There is currently a debate about whether nature should be valued in intrinsic ways or via economic valuation. By looking at the question of *who* values nature in countries in the global south this allows us to explore what might happen in practice when a diverse range of stakeholders are involved such as governments, companies, local communities and indigenous peoples. This chapter will argue that these are important issues that require further research because stakeholders have different agendas (valuation is not a neutral scientific process) and there are often power differentials at play. The chapter features interviews with key academics, consultants and indigenous leaders about how they value nature and their views on monetary and non-monetary valuation. The full interviews can be accessed at: http://whygreeneconomy.org/who-should-value-nature/

There is currently a debate about whether nature should be valued in intrinsic ways or via economic valuation,[1] which in practice often means

[1] Other scholars have also highlighted the importance of relational values alongside intrinsic and instrumental values; see Chan et al. [1] and Arias-Arévalo et al. [2].

D. Kenner (✉)
Anglia Ruskin University, Cambridge, UK
e-mail: Dario.Kenner@anglia.ac.uk

V. Anderson (ed.), *Debating Nature's Value*,
https://doi.org/10.1007/978-3-319-99244-0_10

placing monetary value on "natural capital".[2] The chapters in this book present a fuller discussion on the merits and pitfalls of this approach [4]. This chapter explores a complementary issue which discusses *who* should value nature. It is important to explore this further because valuation is not a neutral process and there are tensions between different stakeholders. I conducted eight interviews in 2014 with academics, consultants and indigenous leaders to find out how they valued nature and their views on monetary and non-monetary valuation. I refer to these interviews throughout this chapter.

The latest assessments by the Intergovernmental Science-Policy Platform on Biodiversity and Ecosystem Services (IPBES) show that biodiversity is declining rapidly in every region in the world. It highlights that biodiversity is under threat from the overexploitation and unsustainable use of natural resources and related air, land and water pollution [5, 6]. In discussing the issue of *who* should value nature I focus on countries in the global south. It is estimated that the global commons of forests, wetlands and rangelands in developing countries support around two billion people and hold a large amount of the world's biodiversity [7]. It is also where areas rich in biodiversity are under enormous pressure to be degraded for oil, gas, mining, logging, livestock, palm oil plantations and large-scale hydroelectric projects. This pressure is likely to increase given that the vast majority of the world's population lives in developing countries [8], many of these countries' economies are reliant on extractive industries and countries in the global north have outsourced resource use [9].

The question of *who* should value nature is pertinent because land rights in the global south in these areas of biodiversity are often heavily contested between states, private companies, indigenous peoples and local communities [7]. In a 2013 report the World Bank estimated that 90% of Africa's rural land remains under customary tenure, meaning it often lacks legal recognition [10]. Throughout the global south mineral, gas, oil,

[2] Natural capital is the world's stock of natural assets including geology, soil, air, water and all living things. Richard Mattison of Trucost consultancy explains the logic: "Valuing the services that ecosystems provide to the economy allows companies and governments to understand the reliance we have on nature and develop strategies to mitigate the economic consequences of environmental degradation. Natural capital accounting is an important tool for companies to measure, manage and reduce environmental impacts. It allows environmental performance to be fully integrated into management processes alongside more established business performance indicators" [3].

forest and agricultural concessions often directly overlap with indigenous peoples and local communities' territories [11] leading to conflicts and an increase in the killings of environmental defenders who fight to protect their lands [12]. Indigenous peoples and local communities legally own just one-fifth of the lands that are rightfully theirs, and this is a particular problem in the global south [13].

Diversity of Valuation

There are several ways in which nature is valued from the spiritual to estimates of economic value [14, p. 11] which are reflected in my interviews. Pavan Sukhdev[3] told me that "[d]ifferent agents value nature's services differently, and using different forms of valuation. We can only be certain that all these valuations will be different." To date it has mainly been specialist companies, consultancies, academia and conservation non-governmental organisations (NGOs) that are valuing natural capital in developing countries [3]. These actors usually focus on monetary valuation because there is probably also a tendency to believe that monetary values will make a stronger business case to protect the environment [3]. These monetary values are also what enable certain actors to encourage investment in natural capital [15]. In some cases this occurs so that it can be traded using blockchain as a verification mechanism [16].

Giles Atkinson[4] acknowledges that "local communities are perhaps less likely to use monetary estimates of the value of nature, giving greater weight instead to other cultural aspects that currently do not lend themselves easily to quantification". Indigenous peoples have a different conception of human–nature relations, seeing nature as a gift [17, p. 9]. Actors such as the Dongria Kondh indigenous community place a spiritual value on the Niyamgiri hills in India referring to the hills as their God and soul (Survival International) [32]. This perspective does not use complex economic models to recognise monetary values.

[3] Pavan Sukhdev is the CEO of GIST Advisory and a United Nations Environment Programme Goodwill Ambassador. He was also study leader for The Economics of Ecosystems and Biodiversity (TEEB). The research by the TEEB project is recognised as one of the key global initiatives advocating for the economic valuation of nature.

[4] Professor Giles Atkinson is a member of the Natural Capital Committee in the United Kingdom which was set up in 2012 as an independent advisory board to the UK government.

My interview with Davi Kopenawa Yanomami[5] helps explain why there are a number of indigenous groups that strongly oppose the natural capital approach [18]. He argues as follows:

> Nature belongs to indigenous and non-indigenous peoples. The people who should value the forest are indigenous peoples because they have a long-held understanding and knowledge of the forest. We are the owners of the forest. It is crucial for indigenous peoples to be able to use nature's wealth. We are the ones who should use the fruits of the forest along with parrots, toucans, monkeys and other animals. Today, the white man is crazy for money. The white man destroys nature for money because he does not value or respect the environment. It is important for everyone who uses nature to understand the spiritual importance of trees, rivers, lakes etc. Today, many people are talking about climate change. The earth is getting hotter. There needs to be a reduction in pollution. I call on non-indigenous peoples to listen to us and learn from us.

It is vital to understand how other actors value nature because indigenous and community lands make up over 50% of the land on the planet [13].

To what extent have these different perspectives been recognised in research? Pascual et al. argue that "nature and its contributions to a good quality of life are often perceived and valued by people in starkly different and often conflicting ways", meaning it is vital to acknowledge a diversity of values [17, p. 8]. They advocate for a *Nature's Contributions to People* (NCP) approach, which has been worked on within the context of the IPBES [19]. Pascual et al. define NCP as "all the positive contributions, or benefits, and occasionally negative contributions, losses or detriments, that people obtain from nature" [17, p. 9]. In a recently published article Díaz et al. [20, 21] make the case that the NCP goes beyond the ecosystem services[6] concept by adopting an interdisciplinary approach that

[5] Davi Kopenawa Yanomami is a leading shaman and spokesman for the Yanomami people who live in the rainforests of northern Brazil and southern Venezuela. He has fought to protect Yanomami lands from illegal gold mining, among other threats, for more than 35 years and has been called the "Dalai Lama of the rainforest".

[6] Ecosystem services are the flows of supporting, provisioning, regulating and cultural services. Services that humans receive from nature include forests that provide timber and river basins that provide fish, ecosystems in the Amazon rainforest that regulate rainfall and capture carbon dioxide, and ecosystems that support and maintain fertile soils for agriculture. Some scholars argue that ecosystem services are co-produced; see [22, p. 246].

recognises the central role of culture and a greater plurality of perspectives, such as indigenous and local knowledge. They argue this could better inform policymaking and decision-making.

However, other scholars have questioned the degree to which this is a new approach as they argue that the diversity of values was already acknowledged.[7] Braat argues that the ecosystem services literature does not have a narrow ecological focus that tends towards monetary valuation. Instead he argues there is a substantial body of ecosystem services research that already factors in cultural ecosystem services (including "a small but increasing number of publications on indigenous knowledge") and social issues [23]. Meanwhile, Maes et al. argue that within the ecosystem approach "there is place for plural values of biodiversity and ecosystems" [24]. They highlight that the European Union (EU) initiative on Mapping and Assessment of Ecosystems and their Services "addressed multiple values in its conceptual model". However, they do acknowledge that "there is less evidence that knowledge of indigenous people is sufficiently taken up in regional ecosystem assessments" and therefore agree that "additional action is needed to include their perspectives in defining the relations between nature and people".

Peterson et al. are supportive of the goal of Díaz et al. to include diverse ways to value nature, with a specific recognition of the importance of cultural aspects [33]. They highlight that there is a "need for a richer process of articulation, translation, and discussion among many different perspectives on people's relationship with nature". But they also note that "multiple values are already being included in ecosystem services assessments". Peterson et al. conclude that the debate around the Díaz et al. [20, 21] article "highlights the importance of including diverse and less-represented knowledge systems in assessments and syntheses such as those conducted by IPBES". They note that the "IPBES was created as a science-policy platform for biodiversity and ecosystem services, and as such it has substantial potential to support and enable these different types of bridging processes" (see more on the role of IPBES later).

[7] For an overview of this debate and how to participate, see Ecosystem Service Partnership webpage https://www.es-partnership.org/ongoing-discussion-on-the-science-publication-assessing-natures-contributions-to-people-diaz-et-al-2018/#more-9256

In reply to the various responses to their article Díaz et al. explained that their intention was to use the NCP framing to build on (not replace) the ecosystem services and open up this debate to include a diversity of other perspectives including the social sciences, humanities and indigenous and local knowledge. They particularly emphasised the importance of indigenous views given that indigenous peoples and local communities "manage a large proportion of our planet's high biodiversity areas" [20, 21 eLetter].

My initial motivation behind asking the question *who* should value nature was to highlight the diverse ways to value nature. There is clearly an active debate around this as the Díaz et al. [20, 21] article and responses illustrate. However, when I began my research I also wanted to go further than this to explore the practical consequences of these different ways of valuing nature. A first step is to look at why stakeholders value differently. In the final section I discuss the importance of power in deciding which valuation is adopted.

Different Agendas

Why do stakeholders value in different ways? As my interviews illustrate valuation is not a neutral process which can sometimes lead to tensions between different stakeholders. Several of the people I interviewed expressed concern about who would be involved. Thabit Jacob[8] comments: "Different actors will value nature with diverse motives. Companies will value nature with the motive to make super profits inspired by greedy nature of the corporate world. Indigenous people will consider intrinsic value of their relationship with nature based on spiritual values which are priceless, and they will resist the idea of monetizing nature." The fact that different stakeholders value differently raises fundamental issues, including around the difficulty of carrying out monetary valuation if there is disagreement over how to value nature.

Thabit Jacob goes on to say that economic valuation "will encourage commodification of natural resources and give much control of such resources to corporations and rich members of the society at the expenses of poor. The process will make more sense if indigenous communities and

[8] Thabit Jacob teaches at the University of Dodoma in Tanzania which hosted the Green Economy in the South conference in July 2014.

local institutions take a leading role in valuation, I believe most will reject". He concludes: "I'm against the recent practice where only private consultants (economists and financial experts) are more involved in nature valuation. Members of academia can also play a significant role in valuation as opposed to consultants, representatives of international NGOs and corporate agencies. My only worry is that some academics can easily be bought by corporations."

Teresa Pérez[9] criticises the argument that it is necessary to place monetary values on nature as a justification to protect it: "Business case for whom? For the communities whose businesses and economies are routinely destroyed by those who now suggest that valuation of Nature might help respect Nature and communities? Or a business case for those who find it increasingly difficult to find sectors to invest with the guarantee of double-digit returns without conflicts with communities?"

She also contends that "[t]he process of valuation is intrinsically linked with the tools that such valuation will be used in. Some argue that the process of valuation is separate from the tool of pricing, yet history clearly shows that the development of the methods is shaped by the tools to be used and vice-versa".

Joan Carling[10] says that "all stakeholders should be involved, and the different perspective and views of nature shall be accounted for and respected, not just the monetary and commercial values of nature. The use, management, utilisation and conservation of nature should not be put in the hands of corporations and corrupt states but should be in the hands of peoples who know the real value of nature beyond monetary terms". It is for similar reasons that Yoni Rivas[11] argues that "[f]armers and indigenous peoples should be involved in valuation. Private banks, companies and international organisations like the World Bank and IMF should not".

[9] Teresa Pérez works for the World Rainforest Movement which was established in 1986 to work with local communities to defend their rights over their forests and territories. It is based in Uruguay.

[10] Joan Carling is former Secretary-General Asian Indigenous Peoples Pact and member of the United Nations Permanent Forum on Indigenous Issues. Founded in 1988 the Asian Indigenous Peoples Pact has 47 members from 14 countries in Asia. Based in Thailand.

[11] Yoni Rivas represents the Movement of Unified Farmers of Aguán (MUCA), which is a coalition working to defend land and human rights in the Aguán valley in Honduras that has seen a number of violent murders of farmers in recent years.

Advocates of economic valuation do acknowledge these tensions. Pavan Sukhdev argues that "[t]he 'business case' for nature is the business case for Enterprise Earth. It is about public goods, public wealth. But private capital only pursues private profits, not public wealth, and that is why we must be very careful how we design and deliver solutions to the problems caused by the economic invisibility of nature".

Giles Atkinson comments: "The important thing is that values do not reflect special interests to the exclusion of other parties who should count in decision making". He goes on to say that "[i]t is important to appreciate that valuing nature (or any other goods and services that are not traded in markets) is not simply about 'assigning' a value or discovering 'the' value of some change in it. Economic valuation is about trying to understand and quantify preferences for environmental changes versus something else that may enhance (or reduce) wellbeing. Ideally, this would reflect the values held by everyone affected by the goods and services provided by nature. This could be all those citizens within a country. Plausibly it could be people living elsewhere too".

Meanwhile Richard Mattison[12] makes the point that "[w]e need a collaborative process that includes all stakeholders. Governments need to conduct natural capital assessments of their ecosystem services and implement environmental policy measures that minimise environmental degradation and reward preservation of natural capital. Businesses need to reduce natural capital impacts in absolute terms, investors need to engage with company boards to ensure environmental risks and opportunities are being managed effectively".

The Power to Decide How Nature Is Valued

Pascual et al. conclude that these different ways to value can lead to conflicts about how nature is perceived, which can "interfere with effective and equitable decisions" [17, p. 8, 25, p. 164]. If there are such different ways to value nature, based on conflicting agendas, how will nature be valued?

[12] Richard Mattison is Chief Executive Officer at Trucost. Trucost was setup in 2000 and is a leading consultancy firm working with companies, investors, governments, academics and thought leaders to understand the economic consequences of natural capital dependency.

Pavan Sukhdev says that "valuation is a human institution, and 'who values', or who is the agent of valuation, is a vital question. From a human rights perspective, it is those who are closest and most dependent on those ecosystems whose valuation matters most. However, provincial and national governments may argue otherwise from an economic or governance perspective. So it is for society at large to decide whose valuation counts".

Exploring who should value nature in the real world context of developing countries reveals that contested land rights in biodiverse regions mean there could be different valuations (monetary or non-monetary) by diverse sets of stakeholders for the same ecosystems. It matters which stakeholders are valuing an area and what methodology they choose to use because this will influence whether they place monetary or non-monetary values on this area. This raises questions about whose values will carry more weight. Do certain stakeholders have the power to limit valuation to non-monetary values or to push this towards monetary values? Who has the power to decide which value is placed on these biodiverse areas? Who will decide whether a valuation is correct or incorrect?

Perhaps there are lessons to be learnt from how ecosystem services are prioritised. Control and access over ecosystem services can influence which services are ranked as more important than others. This matters as there are often inherent trade-offs leading to winners and losers, with some groups benefiting at the expense of others [22, p. 263, 265, 26, p. 576, 14, p. 12, 27]. The most powerful actors often have the most access compared to weaker stakeholders who are limited to the remaining non-excludable and non-rival ecosystem services such as cultural services [28, p. 2, 12, 29]. The unequal distribution of power can also result in certain actors capturing more of the benefits from ecosystem services whilst environmental degradation is pushed on to weaker actors. For this reason some scholars have argued it is more accurate to distinguish between ecosystem services and ecosystem benefits enjoyed by the winners [30, p. 138].

In trying to resolve these tensions, it will be crucial to acknowledge these diverse views to "ensure that value estimates are accepted by the communities they affect" [14, p. 11]. Otherwise, there could be resistance from local communities that feel unrepresented. Pascual et al. argue it is crucial to acknowledge and address "power relationships across stakeholder groups that hold different values on human nature relations" [17, p. 9]. This can be the start of a process towards conflict resolution over values based on "participatory negotiation among stakeholders holding incommensurable values over human–nature relations" in order to "effectively and equitably bridge different value systems" [17, p. 11]. Part of this could entail recognising different types of knowledge, including indigenous knowledge, that could be used to constructively inform ecosystem valuation [31].

Pascual et al. recommend a five-step process as a practical approach to pluralistic valuation. Firstly, clarify the purpose of the valuation. Secondly, choose an appropriate approach that reflects the diversity of values and worldviews. Thirdly, undertake the valuation based on this prior assessment whilst factoring in power relations and trade-offs between values. Fourthly, use transparent and participatory processes to bridge the diversity of values including integrated modelling, multi-criteria decision analysis, narrative approaches, as well as deliberative approaches to give an opportunity for dialogue among stakeholders. Finally, communicate the valuation to the public and decision-makers as part of an ongoing process of reflection [17, pp. 12–13].

In a positive step forward the IPBES has recently announced that in 2018 it will begin a "methodological assessment into the diverse conceptualization of multiple values of nature and its benefits" [5, 6]. The need for this study is based on the recognition that there are "inherent differences between valuation approaches considering different world views and knowledge systems" and that "at present, the design of governance, institutions and policies rarely takes into account the diverse conceptualization of multiple values of nature and its benefits to people". The study will assess different valuation methodologies and approaches based on biophysical, social and cultural, economic, health and holistic (including indigenous and local community-based) values. It will also consider the "perspective of different genders and generations". Crucially this work will be carried out by a multidisciplinary group of experts from different academic disciplines as well "as stakeholders and practitioners relevant to biodiversity and ecosystem services decisions" such as government, indigenous peoples and the private sector.

In what could be seen as adopting a pluralistic valuation approach there is a specific focus on "approaches to the integration and bridging of different types of values" with the potential to develop tools and methodologies that incorporate a mix of values which could be used in decision-making by a range of actors such as government, private sector, civil society organisations, and indigenous peoples and local communities. One of the specific aims is to consider "how different methods and approaches help in acknowledging and dealing with potential conflicts, synergies and trade-offs between the values of different aspects of nature to different stakeholders and sectors".

To date the debate on valuing nature for its intrinsic value or as natural capital has mainly been in abstract terms. By looking at the question of *who* values nature in countries in the global south this allows us to move from the abstract to explore what might happen in practice when a diverse range of stakeholders are involved such as governments, companies, local communities and indigenous peoples. As this chapter has shown these are important issues that require further research because stakeholders have different agendas (valuation is not a neutral scientific process) and there are often power differentials at play.

References

1. Chan, K. et al. 2016. Why protect nature? Rethinking values and the environment. *Proceedings of the National Academy of Sciences* 113 (6): 1462–1465. Retrieved from: https://doi.org/10.1073/pnas.1525002113.
2. Arias-Arévalo, P., Martín-López, B., and Gómez-Baggethun, E. 2017. Exploring intrinsic, instrumental, and relational values for sustainable management of social-ecological systems. *Ecology and Society* 22 (4): 43. Retrieved from: https://doi.org/10.5751/ES-09812-220443.
3. Kenner, D. 2014. Who should value nature? The Institute of Chartered Accountants in England and Wales. Retrieved from: https://www.icaew.com/-/media/corporate/files/technical/sustainability/tecpln13444-who-should-value-nature-web.ashx?la=en
4. Great Transition Initiative. 2014. Monetizing nature: Taking precaution on a slippery slope. Great Transition Initiative. Retrieved from: http://greattransition.org/publication/monetizing-nature-taking-precaution-on-a-slippery-slope
5. Intergovernmental Science-Policy Platform on Biodiversity and Ecosystem Services. 2018. Biodiversity and nature's contributions continue dangerous decline, Scientists Warn. IPBES. Retrieved from: https://www.ipbes.net/news/biodiversity-nature%E2%80%99s-contributions-continue-%C2%A0dangerous-decline-scientists-warn
6. Intergovernmental Science-Policy Platform on Biodiversity and Ecosystem Services. 2018. Information on the scoping for the methodological assessment regarding the diverse conceptualization of multiple values of nature and its benefits, including biodiversity and ecosystem services (deliverable 3 (d)).

IPBES. Retrieved from: https://www.ipbes.net/system/tdf/ipbes-6-inf-9_en.pdf?file=1&type=node&id=16522
7. International Institute for Environment and Development. 2013. "Land grabbing": Is conservation part of the problem or the solution? IIED. Retrieved from: http://pubs.iied.org/17166IIED/
8. United Nations. 2017. World population prospects: The 2017 revision. UN. June. Retrieved from: https://www.un.org/development/desa/publications/world-population-prospects-the-2017-revision.html
9. World Wildlife Fund. 2012. Living planet report. WWF. Retrieved from: http://assets.wwf.org.uk/downloads/lpr_2012_summary_booklet_final_7may2012.pdf
10. World Bank Group. 2013. Securing Africa's land for shared prosperity: A program to scale up reforms and investments. Retrieved from: http://documents.worldbank.org/curated/en/732661468191967924/Securing-Africas-land-for-shared-prosperity-a-program-to-scale-up-reforms-and-investments
11. Rights and Resources. 2014. Communities as counterparties: Preliminary review of concessions and conflict in emerging and frontier market concessions. Rights and Resources. Retrieved from: http://rightsandresources.org/wp-content/uploads/Communities-as-Counterparties-FINAL_Oct-21.pdf
12. Global Witness. 2018. New data reveals 197 land and environmental defenders murdered in 2017. Global Witness. Retrieved from: https://www.globalwitness.org/en-gb/blog/new-data-reveals-197-land-and-environmental-defenders-murdered-2017/
13. Rights and Resources. 2016. Common ground: Securing land rights and safeguarding the earth. Retrieved from: http://rightsandresources.org/wp-content/uploads/2016/04/Global-Call-to-Action_Common-Ground_Land-Rights_April-2-16_English.pdf
14. Bullock, C. 2017. Nature's values: From intrinsic to instrumental. A review of values and valuation methodologies in the context of ecosystem services and natural capital. *National Economic and Social Council* (10). Retrieved from: http://edepositireland.ie/handle/2262/82055
15. Hamrick, K. 2016. State of private investment in conservation 2016: A landscape assessment of an emerging market. NatureVest. Retrieved from: http://www.naturevesttnc.org/pdf/Investing-in-Conservation-2016-full.pdf
16. Sullivan, S. 2018. Nature 3.0 – Will blockchain technology and cryptocurrencies save the planet? Entitle Blog. Retrieved from: https://entitleblog.org/2018/02/01/nature-3-0-will-blockchain-technology-and-cryptocurrencies-save-the-planet/
17. Pascual, U. et al. 2017. Valuing nature's contributions to people: The IPBES approach. *Current Opinion in Environmental Sustainability* 26–27: 7–16. Retrieved from: https://doi.org/10.1016/j.cosust.2016.12.006.

18. Kari-Oca 2 Declaration. 2012. Indigenous peoples global conference on Rio+20 and Mother Earth. Kari-Oca 2. Retrieved from: http://villageearth.org/wp-content/uploads/2012/09/DECLARATION-of-KARI-OCA-2-Eng.pdf
19. Díaz, S., et al. 2015. The IPBES conceptual framework – Connecting nature and people. *Current Opinion in Environmental Sustainability* 26–27: 7–16. Retrieved from: https://doi.org/10.1016/j.cosust.2014.11.002.
20. Díaz, S., et al. 2018. Assessing nature's contributions to people. *Science* 6373 (359): 270–272. Retrieved from: http://science.sciencemag.org/content/359/6373/270
21. Díaz, S., et al. 2018. RE: There is more to nature's contributions to people than ecosystem services – A response to de Groot et al. *Science* 6373 (359): 270–272. Retrieved from: http://science.sciencemag.org/content/359/6373/270/tab-e-letters
22. Palomo, I., 2016. Chapter six – Disentangling the pathways and effects of ecosystem service co-production. *Advances in Ecological Research* 54: 245–283. Retrieved from: https://doi.org/10.1016/bs.aecr.2015.09.003.
23. Braat, L. 2018. Five reasons why the science publication "assessing nature's contributions to people" (Díaz et al. 2018) would not have been accepted in ecosystem services. Ecosystem Services, in press. Retrieved from: https://doi.org/10.1016/j.ecoser.2018.02.002
24. Maes, J., Burkhard, B., and Geneletti, D. 2018. Ecosystem services are inclusive and deliver multiple values. A comment on the concept of nature's contributions to people. *One Ecosystem*: e24720. Retrieved from: https://doi.org/10.3897/oneeco.3.e24720.
25. Garmendia, E., and U. Pascual. 2013. *The justices and injustices of ecosystem services.* Ed. T. Sikor, 161–186. London: Earthscan.
26. Turkelboom, F. et al. 2018. When we cannot have it all: Ecosystem services trade-offs in the context of spatial planning. *Ecosystem Services* 29: 566–578. Retrieved from: https://doi.org/10.1016/j.ecoser.2017.10.011.
27. Sikor, T., ed. 2013. *The justices and injustices of ecosystem services.* London: Earthscan.
28. Felipe-Lucia, M. et al. 2015. Ecosystem services flows: Why stakeholders' power relationships matter. *PLoS One* 10 (7): e0132232. Retrieved from: https://doi.org/10.1371/journal.pone.0132232.
29. Ward, C., Stringer, L., and Holmes, G. 2018. Changing governance, changing inequalities: Protected area co-management and access to forest ecosystem services: A Madagascar case study (30): 137–148. Retrieved from: https://doi.org/10.1016/j.ecoser.2018.01.014.
30. Berbés-Blázquez, M., González, J., and Pascual, U. 2016. Towards an ecosystem services approach that addresses social power relations. *Current Opinion*

in Environmental Sustainability (19): 134–143. Retrieved from: https://doi.org/10.1016/j.cosust.2016.02.003.
31. Sangha, Kamaljit, and Russell-Smith, J. 2017. Towards an indigenous ecosystem services valuation framework: A north Australian example. *Conservation and Society* 15 (3): 255–269. Retrieved from: http://www.conservationandsociety.org/text.asp?2017/15/3/255/215822
32. Survival International. Tribes: Dongria. Survival International. Retrieved from: https://www.survivalinternational.org/tribes/dongria
33. Peterson, G.D., Z.V. Harmackova, M. Meacham, C. Queiroz, A. Jiménez Aceituno, J.J. Kuiper, K. Malmborg, N.E. Sitas, and E.M. Bennett. 2018. Welcoming different perspectives in IPBES: "Nature's contributions to people" and "Ecosystem services". *Ecology and Society* 23 (1): 39. https://doi.org/10.5751/ES-10134-230139.

CHAPTER 11

'Natural Capital': Ontology or Analogy?

Jenneth Parker

Abstract This critical review will initially consider questions of time lags in policy innovation with reference to sustainability. The cultural genealogy of the concept of 'Natural Capital' will then be provided, a key reference point being the 'Five Capitals' approach in Jonathon Porrit's 2005 book. I will suggest that what we have in the Five Capitals approach is a series of analogies conceived as a piece of social marketing of sustainability, primarily aimed at the business sector. I then consider some of the consequences of utilising the concept of 'capital'. Centrally, does this conceptual strategy illegitimately morph into an ideological 'naturalisation' of 'capital' as an essential feature of the world? Does this construction support 'capital' as an ontology that claims universal status rather than a time-bound construction tied to a particular mode of society and historical period? The final arguments concern the question of appropriate transitional concepts for our current situation. What criteria might we apply to consider if a concept or framing is a good one to assist in transition? I conclude that 'natural capital' and the other analogical uses of the term import serious problems. Use with care, if at all: move beyond wherever possible.

This review has greatly benefitted from discussions that took place at the concluding conference in the 'Debating Nature's Value' project. This

J. Parker (✉)
Schumacher Institute, Bristol, UK
e-mail: jenneth@schumacherinstitute.org.uk

V. Anderson (ed.), *Debating Nature's Value*,
https://doi.org/10.1007/978-3-319-99244-0_11

event was distinguished by much real debate and exchange and I thank the varied participants whose points enriched my understanding, including those of editor Victor Anderson.

This chapter is informed by the following: the view that issues of equality and social justice have to be addressed in conjunction with ecological and resource issues for us to have any chance of mobilising transformation required for sustainability; the necessity for a systems approach in addressing the complexity of transitions to a more sustainable world; and an understanding of complexity that includes the need for non-reductive interdisciplinarity, a critical realist approach to recognising key differences in principle between different aspects of the world and our knowledge about them. This chapter faces two ways: on the one hand, it looks back with some empathy and appreciation at the historical period in which 'natural capital' was seen as a useful innovation; on the other hand, it argues for the need to now push way beyond this concept and its limitations, in the current context, and to help enable further moves for a sustainable future.

The Policy Conundrum: Innovation and Time Lags

Understanding the time lags in complex systems is vital for appreciating the sustainability challenge. Time lags underline the need for all the modalities of precautionary action, mitigation and adaptation of negative symptoms that begin to manifest. For example, we know that time lags in the climate system mean that we are now locked into a period of climate change which will get worse before it gets better. Whether it will ever get better is now down to collective human agency at all levels. A key insight at the level of human technical and social innovation is that there is always a time lag between both technical and policy innovation and its application. For example, it is well known in all procurement processes that decisions have to be made for infrastructure investment on the basis of existing technologies and technology assessment processes. Indeed, there is no point in commissioning something that no one currently has the capacity to produce in the amount or numbers required.

Inevitably, newer and possibly more appropriate technologies are developed at some point in the process, and often when the process is committed by having laid the groundwork for the initial choice. There are (at least) two aspects of complexity and potential disagreement here. Firstly, there may be genuine disagreement about what considerations should lead to a decision of 'appropriacy'. For example, should we be considering human user welfare as part of the calculation or not? Equally, on what time scales are we assessing the technology? What kind of benefits do we expect to gain from this technology? Who will gain them? Secondly, we may disagree about the extent to which a certain technology decision will trap us into continuing to support this way into the future, cutting down on our future decision space and flexibility. Resilience and flexibility are now beginning to be much more appreciated as essential in a rapidly changing world—we need to be able to change our minds in the light of changed circumstances without incurring massive costs.

My argument here is that we have to leverage the now existing policy 'embeddedness' of this concept for as much as possible of environmental protection, together with other complementary policy tools, such as the Welsh Government Future Generations Act. However, as a researcher concerned with knowledge for the future, my concern is to point to the need to greatly extend our continuing efforts to move way beyond this concept and its attendant trappings. As a participant in innovative thinking on economy, and from political economy/ecology perspectives, I contend that we are in a much more favourable situation now to change the policy discourse and tools.[1] The global neo-liberal consensus that was manufactured from the 1980s has had a huge crisis, will have more, and its grip is faltering. Furthermore, global awareness of the links between our current model of development and severe environmental degradation has been rising. Now is the time for some real innovation in economy and it is our responsibility to grasp this opportunity.

Aspects of Discourse: Metaphor, Analogy and Ideology

It has been suggested that natural capital is 'just a metaphor', but I suggest that some implications of the discourse we use need unpicking. Should we take discourse more seriously as a field of action, looking at the ways that language and concepts can 'enclose' our minds [3]? Metaphor and analogy are both used in comparisons that we make using language. There is

[1] Some considerations of framing policy in the light of our systems knowledge can be found at http://convergence-alliance.org/download/converge/wp9/WP9_D41.pdf

a fair degree of consensus that 'analogy' is a stronger term than 'metaphor'. In the use of analogy a stronger claim for similarity of two terms is asserted, whereas a metaphor allows more space for individual interpretation. For example, the website for Neurolinguistic Programming (NLP)[2] makes this distinction:

> *Analogy, in its broadest definition is in fact an umbrella term for a cognitive process where we transfer meaning or information from a particular subject to another subject. It is a similarity between the features of two things where a comparison can be made....*
>
> *Metaphor is a type of analogy, but where analogy is identifying two things as similar, a metaphor claims a comparison where there may not be one. It is then up to the listener to create meaning out of this comparison.*

In this respect, I argue that to claim that 'natural capital' is a metaphor is to really understate the case and the claims that are being made. It is an important qualitative difference. 'Natural capital' is clearly intended to be a strong analogy as other structural features of financial capital are asserted as meaningfully capturing features of other areas of life. It is but then a small step to asserting that it is these features that really matter.

Metaphor, Language and Value

In the context of communicating value, metaphor is used in the visual arts, music and drama allowing for its suggestive reach. Many researchers have found that the key thing about the areas of life that people most value is precisely that they are not instrumental 'capital' to be exploited to any other ends, but are seen as ends in themselves [4, 6]. Cultural activities, self-expression, relationships with valued others—these are the kinds of things that people consider transcend the realm of narrow self-satisfaction and can only be cheapened by being spoken about in the transactional terms of the market. Those who are fighting against the creeping dominance of market discourses in the cultural sphere as a part of raising sustainability awareness may well regard their efforts as being undermined by

[2] https://www.networks.nhs.uk/nhs-networks/nlp-in-healthcare/news/metaphor-analogy-and-simile-the-difference-and-why-it-is-important accessed 03.05.2018.

those who are recommending placing everything of human value within a universe of market discourse [8].

Ideology and Naturalisation

Since the introduction of 'natural law' in Christian theology, a key move in developing ideologies has been to claim that the position advocated is, in fact, 'natural'[3] and has always been a feature of the world, even if we have only just discovered this 'fact'. 'Naturalisation' is always opposed to a more historically informed view, supported, one should add, by evolutionary science. Ellen Woods (2015), for example, has emphasised the history of this particular approach to human development, providing a detailed account of capitalism as a very specific kind of market economy. In terms of human history, capitalism is a short experiment and neo-liberalism even shorter. However, the effects of the exponential growth phase of neo-liberalism threaten to derail the whole human experiment, not forgetting the rest of evolved life.

In this context, the formulation of 'natural capital' serves to collude with a view of capitalism as expressing eternal verities about the human condition, and even of life itself. This retro-fitted construction is serving to ideologically prop up the relevance and legitimacy of the neo-liberal attempt to colonise all other disciplines in economic terms set by this programme. Since Margaret Thatcher slammed Hayek down on the table in a cabinet meeting, announcing 'this is our Bible', neo-liberal ideological tools increasingly dominated global development. With regard to 'natural capital' and other kind of ersatz 'capital', we should seriously consider that short-term strategic expediency might seriously harm our ability to shift theory, discourse and practice to a more sustainable system, based on the appreciation of wider qualities of life.

[3] This is why 'natural' discourses are a double-edged sword for environmental movements. Basing our claims in the systems sciences allows space for recognition of evolution and of spiritual dimensions of connectivity and the celebration of systems wisdom of indigenous societies.

Development of the Five Capitals Model

Looking back, how was the concept of 'natural capital' developed and what might its implications be moving forward in this new context of opportunity? Jonathon Porritt's 2007 book *Capitalism as if the World Matters*[4] was welcomed by mainstream liberal theorists such as Will Hutton, editor of the influential *Observer* newspaper. The ideas in this book framed, and continue to frame, research and interventions in the name of sustainable development, including those funded by public money.

Scoones' (1998) original work for DfID on Five Capitals was intended to help describe the livelihoods of poor people in order to help development projects at that time. Whilst Porritt does not reference this work, he clearly builds on it. In Porritt's hands the Five Capitals approach becomes more of a universal prescription for social change: 'if you use this model you will transform your practice in ways that can help lead to sustainability'. There are some genuine differences with Scoones' formulation. Porritt does maintain that 'natural capital' is 'primary' as it comprises the material resources necessary for life. This position is opposed to those who take the view that these different 'capitals' can be inter-substitutable. This position has been further elaborated in diagrams to be found on the current Forum website.[5]

Porritt reconceptualised Scoone's work somewhat proposing the following:

Natural capital—material stocks yielding goods or services
Human capital—health, knowledge, skills, motivation and capacities
Social capital—structures, institutions, networks and relationships
Manufactured capital—material goods contributing to production
Financial capital—no intrinsic value but enables other capitals to be owned and traded

The suggestion is that we should see the world in this way to help us understand both the primacy of natural capital and the need for the

[4] This title was a conscious reference to Fritz Schumacher's *Small Is Beautiful: Economics as if the World Matters* in this sense referencing and claiming links to the Schumacher heritage of radical thought on environment and society.

[5] https://www.forumforthefuture.org/project/five-capitals/overview

maintenance of social capital. However, it is still open to questioning whether the stated aim of the recognition of 'natural capital' as providing the essential bases for all life and economy is best served by the Five Capitals construction. If we were to really prioritise the living systems of the Earth instead of using financial 'capital' as the analogy, we could come up with a very different set of elements.

Porritt is sensitive to the critiques that using the concept of capital to describe the world imports the ideology of commodification. His argument for the validity of this approach is that capitalism is the only game in town and its terminology is so dominant that we must use it in order to communicate a reforming agenda:

> *If there is any sustainable variant of capitalism out there, then it will need to work with the conceptual and linguistic conventions that people are now so familiar with. The concept of capital serves not only to explain the productive power of capitalism; it also provides the clearest means of explaining the conditions for its sustainability.* [7, p. 113]

Key Problems of Using 'Capital' as an Analogy

So what are the key problems in using 'capital'—a concept taken from its context of 'financial capital' to describe the key elements needed for human life?

Fungibility Versus Multiple Goods

One the first things is that 'financial capital' is fungible (9, p 15); this means that, whilst we might have different currencies in the world, financial systems establish equivalence in simple numerical terms, thus allowing for trade. So one quick answer to the problems of the financial capital analogy is that the other proposed 'capitals' are precisely NOT fungible—they are all needed and cannot be inter-traded without judgement and value calls being made.

The concept of 'Five Capitals' therefore creates confusion. Using 'capital' as a guiding analogy does not solve the problem of complex values that need to be politically negotiated; in fact it has the pernicious effect of obscuring these issues. The only vision for addressing the problem of sustainability that this supports is a managerial approach, subject to capital calculations, envisaged as being led by a kind of elite vanguard in cooperation with business. This is a view in opposition to those who propose that sustainability must entail a fundamental political and economic rethink of the nature of the good, indeed, the possible, life. The difference with other previous periods of social and political change is that, in the case of sustainability, it is essential that the sustainability sciences inform the debate. This is why forward-looking organisations such as the German Advisory Council for Global Change (GBWU) have proposed a 'new social contract for sustainability'—a new contract between science and civil society.[6] There have been many other moves towards greater civic engagement with the sustainability sciences.

The Extractive Nature of Capital

Failure to engage with the inbuilt extractive meanings of 'capital' undermines the Five Capitals project. We have now accumulated a lot of research and knowledge about the profound unsustainability of the development experiment of the capitalist era. This knowledge demonstrates that, in addition to the extraction of value from human workers, this model is based on extracting value from living cyclical systems (TEEB 2018). This extraction of value, and generation of waste, is continuing to such an extent that we are contemplating the global collapse of biodiversity and potentially irreversible changes in the stable climate that has supported human development thus far. Climate change can be considered a 'waste' problem as we are simply using the atmosphere like a dustbin.

From this perspective, to support accounts of 'natural capital' is to immediately bring in a presumption that living systems can appropriately be subject to an extraction process, without the understanding that the processes that constitute life and life support need to be respected, maintained and regenerated. This is also a profoundly Western ethnocentric viewpoint that fails to take account of the language and perspectives of those actors globally who best understand the cyclical nature of living systems and how to maintain them and gain a living at the same time.[7]

[6] The author has proposed a complementary four-cornered model of 'good governance' to include Civil Society, Business, Government and Sustainability sciences in the essential role of decision support.

[7] It is this reasoning that has led the Food and Agricultural Organization (FAO,) to conclude that small-scale closed-loop farming is the way we should be seeking to feed the planet.

'Capital' Cannot Be an Ontology for Interdisciplinarity

Some have claimed that this approach can usefully frame research proposals and projects on the spurious grounds that it can usefully unite different disciplines:

> *More interdisciplinarity is certainly also needed if one consider sustainable development as a process where every form of productive capital must stay in line with each other. By this we mean manufactured capital of course, and natural one but also human and social capital as well.* [2, 1]

To propose that 'capital' is a term that unproblematically crosses disciplinary boundaries, and can therefore help interdisciplinary research and understanding is unsubstantiated—in fact the opposite is true [11]. This approach is reductive of the knowledge produced by other disciplines and takes us further into the implicit acceptance of 'capital' as the main ontological category in the world. This universalist claim for the concept of 'capital' is in line with the advance of 'economism' into all areas of life [12] under neo-liberalism. Further, it is evidence of the failure of mainstream economics to take seriously the cogency of other areas of knowledge vital for our future—centrally including Earth system science. What is at stake in interdisciplinarity is the provision of a non-reductive but still coherent frame for the interrelation of different aspects of reality [5]. 'Capital' does not do it.

'Transitional' Concepts and Wider Debate on New Economics

We have seen some proponents of the 'natural capital' concept claim that it is a 'transitional' concept. One very important consideration as to whether a concept might be useful in 'transition' is that in order to assess this it is necessary to have some idea where you want to be heading. This is why we need to consider the wider issues of economic transition and our overall goals in the context of evaluating 'natural capital'. This is not the place to fully rehearse all the debates about the possible nature of a sustainable economy; the main concern here is to argue that the key research agenda is in what ways we might reform economics to replace presumptions that are perniciously extractive to life-support systems and biodiversity. It is only in this way that we can develop and advocate for changes that can truly give some hope for sustainability. If we waste our time and resources on approaches that have built-in failure, we are acting irresponsibly.

In addition to many other avenues, such as looking at de-growth and the rise of the social economy, this involves, for example, exploring the possibilities of 'Green Keynsianism'. The Keynsian argument, born out of the failure of austerity, is that businesses will collapse if workers do not have the means to buy goods. The economic feedbacks are so visible that even the World Bank is now making Keynsian statements.[8] By contrast, we know that the collapse of living systems is a chronic and apparently slow-moving catastrophe, often hidden from view. Can we develop a transitional approach that could lead to a more thorough-going reorganisation of our economies and significant changes to capitalism as the globally dominant form of production? Systems approaches to economy carry out this task in a less reductive manner, by seeking to understand the economic drivers and feedbacks of this system and links with other kinds of system.[9]

To be transitional, any strategy needs to resist the canalisation or 'locking-in' of our trajectory, to give us maximum room for manoeuvre as our dynamic situation progresses and changes. If major change is inevitable, then it is also reasonable to demand that any transitional strategy helps lay some foundations for increased resilience to the trauma. Some would argue that if policies based on 'natural capital' help us to argue for the preservation and value of at least some of our still living systems, this will be justification enough. But 'enough' is the key word here. In our current situation this is not 'enough' when the whole system is in need of change.

For those who see 'natural capital' as a useful interim policy strategy, these considerations may appear to be lacking in realism, given the short-term tasks that confront us. However, in the transitional project of sustainability we have to try to find leverage points in our systems that can help us make changes that do not, in turn, trap us into new or continued forms of society, economy, culture and human being that will undermine

[8] The main argument of Yanis Varoufakis is that the global economy lacks a Keynsian wealth recycling system. His excellent book *The Global Minotaur* does not show any awareness of environmental extractive issues, however.

[9] https://www.sedlabanki.is/library/Skraarsafn/Malstofur/Harald%20Sverdrup%20-%20System%20dynamics%20as%20a%20tool%20for%20economic%20analysis.pdf

our longer-term goals. My question to those who are utilising these concepts currently is 'can we use these concepts whilst continuing to appreciate and argue for the deeper changes needed?' In this difficult task it will be important to continually reflect upon the limitations and problems of such analogies, and to be alert to opportunities to move to the next phase of the social and cultural challenge.

Times Have Already Changed

The context within which 'natural capital' was developed was characterised by the continuing dominance of neo-liberalism and was prior to the financial crisis. Whilst economic theorists have attempted to 'dig in' behind the barricades and defend their positions, the world and opinions have been changing around them. There is now a global movement of students demanding to be taught alternative economic theories,[10] there is a great rise in the global de-growth movement, departments of political ecology are becoming centres for debate on new economics, there is an international well-being economy alliance,[11] and so on. The context of the growing inequality and the recognition that 'growth' has simply made the rich richer are contributing to a sense of possibility about rethinking economics. The sustainability movement can only truly progress by greatly increasing strategies to deal with inequality alongside ecological protection and regeneration. This is not the time to be propping up the old regime of extraction and accumulation by a very privileged few. In terms of social marketing it is our job to sell, but also to engage people in developing, a new social and cultural project.

Conclusion

I do not dispute that organisations that engage with business can be doing valuable work, as part of an ecology of activism and movements. The problem arises when an approach developed for this particular constituency is inflated to a claim that the explanatory strategies used can act as a 'framing' for the whole sustainability movement. This is particularly

[10] http://www.post-crasheconomics.com/

[11] http://www.rapidshift.net/a-movement-of-movements-to-transform-to-a-wellbeing-economy-weall/

problematic when the frame advocated coincides so happily with the main terms of the dominant discourse at the root of the problems we are trying to solve.

I argue that the continuing prominence of the analogy 'natural capital' has the ideological effect of helping to keep the sustainability debate within bounds that do not challenge current social relations. This also gives rise to some manifest contradictions with this approach and other claims to support people's agency for change. The Five Capitals approach is that it keeps our thinking within the boundaries of capital used as an analogy, and implicitly inflated to an ontology—or claims about what really exists in the world. The ontology that we do need is one which makes clear our shared human embodied existence on a finite planet, along with a host of other living species and our absolute dependence on Earth systems for the future. This is the time when we need to use our brilliant emergent cultural, intellectual and technological capacities to engage with this challenge.

This piece does not propose a set of ready-made alternative answers—but what it does do is to argue for the priority of other research programmes for rethinking economy. Using 'capital' as an explanatory concept is not in a fix-it solution, but is better regarded as a holding operation, working at mitigation of a dysfunctional system, but failing to challenge its essentials. At the very least, the gathering extent of the global ecological crisis demands a thorough reformation of capitalism, it is this that we should be demanding right now [14].

References

1. Brechet, T., and Boulanger, P.M. 2002. European and global approaches: IDD overview, classification and characteristics of scientific tools for a Sustainable Development policy. In *Proceedings EU Workshop on Development of scientific tools in support of Sustainable Development.*
2. Dasgupta, P., and K.G. Maler. 2001. *Wealth as a criterion for sustainable development.* Stockholm: Beijer International Institute of Ecological Economics.
3. Ford, L., and Kuetting, G. 2017. Global environmental governance in the Anthropocene: Breaking out of the enclosures? *System Change,* [S.l.], 1(1) July. ISSN 2396–7293. Available at: https://systemchange.online/index.php/systemchange/article/view/15. Date Accessed 31 May 2018.
4. Kenter, J.O., N. Jobstvogt, V. Watson, K.N. Irvine, M. Christie, and R. Bryce. 2016. The impact of information, value-deliberation and group-based

decision-making on values for ecosystem services: Integrating deliberative monetary valuation and storytelling. *Ecosystem Services* 21 (Part B): 270–290.
5. Parker, J. 2014. *Critiquing sustainability, changing philosophy.* London: Routledge.
6. Pfeffer, J., and S. DeVoe. 2009. Economic evaluation: The effect of money and economics on attitudes about volunteering. *Journal of Economic Psychology* 30 (3): 500–508.
7. Porritt, J. 2007. *Capitalism as if the world matters.* London: Routledge.
8. Read, R., and M. Scott Cato. 2014. A price for everything: The natural capital controversy. *Journal of Human Rights & Environment* 5 (2): 153–167.
9. Roderick, I., and the CONVERGE Project Team. 2013. Report on how ranking of degree of convergence may be undertaken – calculating a convergence baseline. CONVERGE Deliverable 16. The Schumacher Institute for Sustainable Systems, Bristol, UK. http://convergence-alliance.org/download/converge/wp2/WP2_D16.pdf.
10. Scoones, I. 1998. *Sustainable rural livelihoods: A framework for analysis.* IDS Working Paper 72. Brighton: IDS.
11. Spangenberg, J.H.C. 2016. Blind spots of interdisciplinary collaboration monetising biodiversity: Before calculating the value of nature, reflect on the nature of value. *Trieste* 3 (1): 115–128.
12. Teivainen, T. 2002. *Enter economism, exit politics: Experts, economic policy and the damage to democracy.* London: Zed Books.
13. Varoufakis, Y. 2015. *The global minotaur.* London: Zed Books.
14. Von Weizsacker, E.U., and A. Wijkman. 2017. *Come on!: Capitalism, short-termism, population and the destruction of the planet.* New York: Springer.

CHAPTER 12

'Natural Capital' and the Tragedy of Environmental Value

John Foster

Abstract The concept of natural capital offers environmentalists a dangerous metaphor. While it has limited positive uses for implementing the recognition that natural resources and the capacities of ecological systems are not free goods for the economy, at the level of earth systems it radically misrepresents the relationship between human beings and their planetary home. I argue that this is an economistically inflected instance of a more fundamental misrepresentation, that involved in seeking to attach *values* to our constitutive natural context at all; but also that we are so constituted as not to be able to avoid doing this so that value-conflict becomes tragically inescapable in this domain.

The natural capital concept leads us radically to misrepresent the way in which the natural world matters to human beings. It is not just that there are serious problems with the economic valuation techniques involved; that the resulting monetary values are often artefacts of the measuring process, or at best little more than codified guesswork, has been contended for almost as long as the model has been around, although this is an

J. Foster (✉)
Lancaster University, Lancaster, UK
e-mail: j.foster@lancaster.ac.uk

V. Anderson (ed.), *Debating Nature's Value*,
https://doi.org/10.1007/978-3-319-99244-0_12

objection which I will refrain from pursuing here.[1] The more important point is that anyone not committed for professional or other reasons to making that methodology appear workable will readily admit to strong intuitive doubts about the whole conceptual framework. A construct which runs to estimates of "the value provided by the Earth to the global economy" (US$51 trillion in 2011 prices, according to the World Bank and Robert Costanza[2]), or even to apparently less ambitious claims such as that bees in their pollinating role contributed £650 million to the UK economy in 2012,[3] easily feels as if it just can't be trusted when we are asking what priority we should give to protecting the biosphere and the atmosphere from further anthropogenic derangement. I want to argue both that this intuition is sound and that seeing why it is so should raise serious questions for our understanding of environmental value itself.

A Dangerous Metaphor

Such an argument need not deny the initial tactical appeal of the natural capital idea. Capital is, by definition (see any economics primer), a stock or asset which has the characteristic of producing a flow of income or some other benefit, and which is valued as the net present value of this flow. The notion that not only natural resource stocks but also ecosystems and natural areas in general could helpfully be thought of as capital assets first took hold in the 1970s and was intended to promote conservation, then becoming increasingly topical, by ceasing to treat these stocks and systems in economic modelling as indefinitely available free goods. (For the back-story here, see [1, pp. 8–9]) The essentially metaphorical transference of the capital concept into this sphere offers a prima facie attractive way to give environmental considerations purchase in a culture where economists are taken much too seriously—or, to be more polite, where economic and accounting concerns, the claims of the bottom line, carry very great and sometimes excessive weight. The seasoned environmental campaigner

[1] Those interested should see the papers in Part II of Foster (ed.) [4]. The case has been reiterated, in essentially the same terms, many times over the succeeding two decades.

[2] Quoted by Richard Spencer in "Costing the earth" at https://www.icaew.com/-/media/corporate/files/technical/sustainability/costing-the-earth-oct-13.ashx?la=en (accessed 15.3.18).

[3] See https://www.telegraph.co.uk/news/earth/wildlife/11679210/Bees-contribute-more-to-British-economy-than-Royal-Family.html (accessed 24.4.18).

Tony Juniper has indeed gone so far as to suggest that putting this metaphor to work is a way of "reconnecting people with the reality which we inhabit"[4]; and at the scale of particular choices, where what matters is alerting decision-makers to the fact that natural systems are not just there to soak up externalities but to make powerfully positive contributions to our well-being, this may well be justifiable. To take one of Juniper's own examples from the talk cited, it is easy to overlook the vital importance of the complicated ecosystem which is the ordinary soil beneath our feet. Even the crudest costing of the "services"—water retention, carbon absorption, nutrient recycling—which it provides will serve to emphasise that its purely economic value is likely to exceed significantly the value to be realised through market-priced activities such as concreting over it or compacting and eroding it by destructive industrial farming methods.

But it is one thing to recognise in the capital metaphor this kind of limited tactical usefulness, and quite another to let it effloresce into a mind-virus colonising and vitiating our general approach to the natural world. How this tends to happen, and the dangers which it poses, can be illustrated from the attempt to bring the model to bear on sustainability thinking by distinguishing in its terms between weak and strong sustainability policies.

Weak sustainability is defined as the position that there is no objection to depleting natural capital assets over time so long as these can be substituted for by manufactured capital (e.g. natural biological productivity by genetically modified organisms, or the CO_2-absorptive capacities of the atmosphere by carbon capture and storage systems) in order to maintain future benefit flows at present values (or better). Strong sustainability, by intended contrast, grounds a predisposition against depleting natural capital precisely on the fact that a good deal of it is *not* effectively substitutable by any such human-made alternatives. (See [2, p. 76] for a classic account, and one which Barbier, *op.cit.*:208 essentially recapitulates, so this is still very much a live issue.) Such non-substitutable natural assets are then claimed to constitute, in the terminology which emerged to mark this key distinction, *critical natural capital* [3]. The practical policy difference here is supposed to lie between an approach favouring careful and appropriate development, in the course of which natural capital assets may be

[4] See https://naturalcapitalcoalition.org/tony-juniper-talks-natural-capital-at-harmony-in-food-and-farming-conference-2017-video/ (accessed 24.4.18).

reduced or downgraded provided overall benefits (to human beings) are maintained or enhanced, and one favouring robust protection of at least the critical component of such assets. (The latter approach will also strongly favour precaution, on the basis that ecosystemic complexity entails our often not knowing in advance which natural capital elements *are* critical.)

The trouble is, however, as Alan Holland pointed out over 20 years ago [8], that this attempt to mark a distinction between strong and weak sustainability collapses as soon as any pressure is applied to it. Weak sustainability says that the various forms of capital (and in particular, manufactured and natural capital) may be intersubstituted where this enhances or at least preserves the value of the benefit flows. But strong sustainability, on inspection, says just the same thing with a different emphasis. That we can't preserve benefit flows by substitution where we have as a matter of fact no adequate manufactured substitutes available is not only not incompatible with weak sustainability, it is actually a corollary of it. What the proponents of strong sustainability evidently wanted to capture was the thought that at any rate some of the natural world should be protected *as such*—that there doesn't just happen not to be, but *couldn't* be, any manufactured substitutes for it. But the concept of critical natural capital won't do this job for us. Once we have a capital assets model for the natural world at all, there can be no objection in principle to substitution going through where it can—since you can't consistently take the point of something to be the production of benefits, while denying that it could be substituted for by anything else which would produce the same benefits just as effectively. If critical natural assets are substitutable at least in principle, that is, they seem not to be in the intended sense critical, while if they aren't so substitutable, they aren't capital.

Nor can the desired immunity of critical natural capital from human interference be saved by recourse to Daly's point about complementarity (*op.cit.*:77)—that manufactured capital can only complement, not entirely substitute for, natural systems: as fishing boats, for instance, can increase the capital value of, but evidently not stand as proxy for, the population of fish in the sea. But the flow of value from *any* capital asset will always be a function not just of the actual stock comprising it, but also of the technical means which we employ in using that stock. (If you have two tractors and invest in a better kit of spanners for servicing them, you still have just the two tractors but their capital value will have increased.) Hence if we think of the atmosphere, for instance, as a capital asset, we can hope by various means to assist it in its functions of climate regulation aligned to our purposes, where these include not only having a habitable Earth but also continuing to do at least many of the CO_2-generating things which we want to do—we can hope to substitute various technologies, that is, not (unimaginably) for *the atmosphere*, but for part of its capacity to facilitate human purposes, thereby restoring asset value from which we have derogated by raising levels of atmospheric CO_2. While complementarity of natural and manufactured capital must clearly be the situation *overall*, since the Earth is the stage and source for all our activities, manufactured capital can in principle perfectly well substitute for some functions of natural systems when these are taken as capital assets subserving specific human purposes.

Very importantly, however, this implication of our deploying the capital metaphor doesn't just hold "in principle" —it has significant practical consequences. It means that there is nothing to inhibit a policy of pushing the replacement of natural by manufactured capital as far as our technological ingenuity allows, so long as we ensure—or (which is really the key point) can plausibly *represent* ourselves as ensuring—that substitution and not derogation is what is really going on. The invocation of criticality was supposed not just to underline the point that we have to start from natural capital, but to provide a robust ground for our not further interfering with life-support systems like the atmosphere. Once we start thinking of nature as capital at all, however, it will always be an empirically open question whether we can effectively substitute manufactured capital for those of its key functions in providing future benefit streams with which we are for the moment concerned—and, crucially, under pressure of present needs we will always be disposed to assume answers to that question which will favour our continuing for the time being to degrade the natural systems. So, for instance, we will go on emitting CO_2 vastly in excess of what the 2°C threshold for preventing disastrous global warming would permit, and we will even go on drilling for more carbon-based fuel, while pinning our hopes on the untested possibilities of carbon capture, or nuclear power, or geoengineering the atmosphere with sulphate particles, with all the rich potential of these and similar technologies for unforeseen side effects and planet-wide screwing up (on which see, for instance, [9]). The model of ecosystems and other natural resources as capital assets not only provides us with nothing to restrain these tendencies, but actively incites us to indulge them. We encounter here, in fact, just the same dynamic as informs the general failure of supposed 'sustainability' constraints to constrain—that is, of contracts with the future haggled over only amongst ourselves to have any seriously binding force.[5]

This is all bad enough. But it also begins to expose a difficulty which goes deeper still. This is that, whether or not and to whatever extent supplemented by our technical ingenuity, the background natural affordances which have (hitherto) framed the human habitability of the planet—the suite of predictable seasons, the endurable temperature range, the hydrological cycle yielding plentiful freshwater, the abundance of protein in animal and vegetable forms—surely cannot be thought of as *good for humans*, any more than it can be a good thing for you as an individual human being that you are embodied. You aren't better off embodied than you would be otherwise—to be sure about which you need only ask: than *who* would be otherwise? The natural affordance of human embodiment matters not just *to* you but *as constitutive* of you, and in the same way those long-standing biospheric and atmospheric affordances which make human life possible—our enEarthment, as we might analogously call it—matter as partly constitutive of our species being. A humanly habitable world is one which supports the specifically human form of life—that is, in very general terms, one where the coherence of communities large enough for the development of a rich, complex, linguistically framed sociality is enabled by reliable dependence of local habitation on the surrounding biosphere for its provisioning across successive generations. Without the conditions of environmental and climatic stability on which this form of life relies, we wouldn't be worse off, because *we* wouldn't *be* at all. But any kind of benefit by definition makes its recipient better off; so background natural systems can't be providing humanity with a flow of benefits; so they can't sensibly be modelled as any kind of capital asset. It is not just methodologically flaky, but from this perspective actually *absurd*, to claim any figure as "the value provided by the Earth to the global economy". A natural capital model for these Earth systems seems to make no more sense than it would for you to think of your body as "corporeal capital", something providing an important flow of asset value to support your engagement in football, sex or knitting. We would thus appear to have reached a level at which, so far from reconnecting people with the reality they inhabit, the capital metaphor actually turns the key relationship of

[5] On which, see [5].

dependency between human activities and natural systems completely on its head.

EnEarthment and the Claims of Value

How does an initially useful analogy for registering nature's importance lead us into such an apparently gross misconstrual of this fundamental relationship? Under what pressures are we disposed to let this metaphor run away with our thinking to that extent? It is, I believe, much more than a matter of our being caught up in globalised late-capitalism and its characteristically economistic assumptions, tiresome and humanly impoverishing though all that is. Rather, some basic issues about how things matter to us drive the overextension of this metaphor in this cultural climate. The natural capital model, I want to suggest, represents a particularly baroque and economistically inflected version of a much more general disposition, one on which the whole discipline (or pseudo-discipline) of environmental ethics has been founded. This is our commitment to thinking that we pay proper regard to the natural world in which we are enEarthed by recognising it as the bearer of moral or prudential *values*—prudential insofar as we think of it in relation to human interests, moral insofar as we take aspects of it to deserve respect or care even at the expense of human interests.

Diagnosing this commitment, and indeed seeing that diagnosis is what it calls for, requires us to go back briefly to basics. *Things matter* because of life, and across the web of life. Nothing matters for a stone, but as soon as you have life, the idea of *import* gets a foothold. Rain and sunlight matter for a tree—their presence or absence make a difference to its flourishing. Water and fish matter for an otter, and they also matter *to* it: as a sentient creature it is aware of them, suffers in their absence and can also easily be observed to delight in their presence. The multiplicity of ways in which their environmental circumstances matter, that is, carry life-relevant import, for and to living things is in fact what warrants the image of a *web* when we are thinking of life's ecological interconnectedness—a web, of course, in which we humans too are wholly implicated.

I am indebted for insisting on the fundamental insight here to an original and thought-provoking paper by Tom Greaves and Rupert Read [7]. They, however, want to represent *all* this ecological mattering as *value*—so that there would be a vast deal of value around, and valuing going on, entirely outside the arena of human interests and concerns. (This is as a prelude to arguing that the natural world is pervasively valuable apart from those interests and concerns.) But while the ecological basis is indeed crucial, there is strong reason for limiting the idea of value solely to the human realm. Value is always anthropo*genic*, and although it need not be anthropo*centric* (we certainly need not think that the only things that matter are things that matter to us), it is always indelibly marked with the human signature. Therefore it is best to retain some other term, perhaps *affordance*, to stand for the wider mattering that goes with life in general. And this is not just a terminological quibble: the distinction is important because, while all value belongs inherently within the domain of rational comparison, some affordance, as emphasised earlier, is *constitutive*, and thereby falls outside the scope of comparative importance.

What do I mean by that? The natural affordance of water matters for and to an otter, for example, by virtue of its relation to the truth about the world expressible in the sentence "Otters are aquatic (or semi-aquatic) mammals". Propositions like this, so-called Aristotelian catagoricals [12] which organise our understanding of forms of life, have different logical properties from universally quantified propositions like "All otters live in or near water". (Notably, the latter but not the former is falsified by finding one which is kept, cruelly, in a cage away from water.) A further aspect of this now widely recognised logical difference which reveals itself as relevant to our present context is that while it makes perfectly good sense to say of Oliver the captive otter that *his* state of deprivation matters a lot for and to *him*, and that he would be a lot better off in his natural aquatic state, it is at least infelicitous and arguably nonsensical to say that water *matters a lot*, or even matters *enormously*, to otters as such, as if water was something very high on a comparative scale of things that matter for a species' existence conceived of independently from its necessary life conditions, and so affording them a much better life than they (than *what*?) would have had otherwise. Of course, otters don't themselves evaluate their affordances—they aren't blessed (or cursed) with that kind of conceptual equipment. But insofar as some things, such as easily available fish as against the strenuous exercise of catching elusive ones, might be represented as being of differential import to them, that can only be so from within, as it were, the constitutive import of their natural condition as simply given for their form of life.

Value, however, which is the way in which human beings, reflexively conscious of themselves and their intentionality, *do* characteristically articulate and respond to encountered matterings in the world outside them, always subsists in what the American philosopher Wilfrid Sellars [11] famously called "the logical space of reasons", and operating in that space as a practical agent is defined by one's exercise of the capacity to judge *comparatively*. Any value, to *be* a value, must always be more or less (or equally) pressing, more or less (or equally) a justification for action, by comparison with some other value. John McDowell, a more recent philosopher in the same vein, describes our being initiated into this space of reasons, which happens through the natural course of an individual human being's growth and development, as acquiring a *second nature*, essential to which is "a kind of distancing of the agent from the practical tendencies that are part of what we might call his first nature" [10, p. 188] This distancing is a necessary condition of our entering the space of reasons at all. Reasons are inherently general: if anything can be taken as a reason for me (to believe or act), it must be taken as a reason for *anyone in my situation* to do so; and then, at the same conceptual stroke as it were, I am constituted as a self separable in principle from my whole situation (including of course my environmental situation) and the reasons which it gives me. Within the space of reasons one must always be able, as McDowell puts it, "to step back from *any* motivational impulse one finds oneself subject to, and question its rational credentials" (ibid.), so that when one acts, one acts on the motive which is comparatively *best* credentialed.

But as we have seen in relation to natural capital, a biosphere functioning to provide an independently self-regulating living context for human existence, as it did from when humans emerged right up until the final quarter of the last century, is the source of key natural affordances for our species life which are importantly constitutive of human beings. As such, that self-regulating natural world, un-destabilised by an overdominant human presence, matters to us *incomparably*, as the absolute precondition of any evaluative comparison we may engage in. So the biosphere, by an argument which in effect generalises that about its supposed 'capital asset' status, can't matter to us in the way in which anything which has value matters.

The Tragic Trap

When I say that the biosphere is a precondition of evaluative comparison, it might seem as if all I am really saying is that we have to be alive in order to be able to value anything—which would be true, but pretty banal. I am, however, trying to say something rather more philosophical than that, in the sense that one of philosophy's important roles is to place the apparently banal in an interesting light. The point is that just *because* we have to be alive to value, it is a deep mistake to see the maintenance in being of our necessary life conditions as something which can have a value for us. Rather, the conditions of viable human life are literally *invaluable* because they ground all valuing, just as the visual field is itself literally invisible, and thus they can't intelligibly be weighed, even in principle, against anything else which might matter to us in the way of value. It is a conceptual-grammatical solecism to suppose ourselves able to ask how much emphasis we should put on, say, reducing the driving forces of climate destabilisation *as against* countering global economic injustice or promoting greater gender equality. It is essentially the same kind of mistake as if I wondered how important it was to me to be alive, *as against* being kind or fair. We can see without any prompting why *that's* silly—so why not in the case of background natural systems?

Here I think we reach the root of the matter. As beings with a dual nature, we are tragically liable to be caught between our first nature and our second—to encounter certain key life affordances as constitutive of our species being, but only to register them as mattering to us in the "space of reasons". Correspondingly, while our first nature is wholly permeated by dependence on its biospheric context, our second seems to set us up independently, vis-à-vis anything else which we can rationally consider and evaluate. And the legacy of the Enlightenment in particular works to spring this tragic liability as a *trap*, so that we simply cannot get our heads around anything's being objectively important to us (mattering, that is, as more than just what we happen to desire) except in a way that is accessible to comparative evaluation by the disembedded rational subjects which we also cannot help taking ourselves to be.

If we revert to the natural capital metaphor at this point, we can perhaps see the conceptual machinery of entrapment in operation. It makes no sense, as we noted, to think of the natural affordances which support our specifically human form of social being as benefiting humans as such—that is, human beings as the subject of 'Aristotelian categorical' propositions about the dependence of their form of life on a reliably stable environment. But clearly, derogations from the stability of that environment can constitute, as they are currently doing, very serious disbenefits for—and, correspondingly, partial restitutions of that stability can immediately benefit—the enumerable set of human beings alive at some particular time. Thinking of resources devoted to such restitution as investment in natural capital, for our own or our successors' benefit, therefore, misrepresents the reality of our enEarthment by *systematically occluding* our status as one more life form in a specific relation of dependency on the natural world. All we see, in the grip of the model, are sets of historical individuals confronted with the need to make rational choices about their particular impinging circumstances.

Similarly, at the more general level of value, which 'natural capital' is one attempt to operationalise: within the Enlightenment trap, morally unimpeachable goals like worldwide equality, political and economic justice and so on are pursued by (in a nutshell) trying to globalise the historically accidental material standards of a Western lifestyle built on fossil fuels, in the process massively polluting, degrading and destabilising life-support systems worldwide—and then we ask: how can we identify values for those life systems which will enable us somehow to combine and balance proper respect for them with due regard for material 'progress' and all our other aspirations to human betterment? We have here, in fact, a generalised version of the way in which the natural capital concept betrays us in practice. That model, recall, licences us to rely on ways of allegedly maintaining the capital value of the natural world while actually continuing remorselessly to diminish its extent and resilience. These ways range from the almost insanely hazardous—geoengineering the stratosphere with sulphate particles, for instance, so that we can go on pumping out CO_2—to the grimly comic (the last male white rhino pegs out, but there are still a few females around and we've got his semen in the fridge, so maybe we can still hand on as 'capital' this marvellous beast which humans have driven to near extinction). Thus we let ourselves off the really difficult challenges (stop emitting CO_2, stop robbing life space from the non-human) which ultimately we just have to meet. At the more general level, essentially the same thing is happening: the way we think about all mattering as value systematically occludes the stark claims of the *in*valuable (where that means, not the very, very valuable, but the literally non-evaluable). The kinds of item in relation to which it does make at least prima facie sense to think about comparative importance (this woodland, a particular landscape, the Great Barrier Reef, etc.) get jumbled up with the kind of thing which we desperately, as one might say out of our species being, need to recognise as *beyond value*—an atmosphere with no more than 350 ppm of CO_2e, a self-regenerating biosphere. These then join in with the values of reducing (human) hunger, widening equality and keeping the peace, all clamouring to be attended to as best we can. "The environment" inevitably thereby ends up as one, and by no means always the most pressing, among a range of competing demands on our moral energies. But to place ourselves in this position is *already* to have sold the crucial—the literally vital—pass.

The genuinely tragic nature of this situation must be re-emphasised. Value is simply the wrong kind of conceptual currency in which to deal with constitutive natural affordance and dependency, but it is the only one which seems available to human beings as rational agents, since for such agents there are no absolute givens except rationality itself—only its procedural requirements can be categorical. Hence, we cannot help disembedding ourselves in the very process of trying to grasp our constitutive embeddedness, or seek to manage and control for our rationally chosen ends the naturally given preconditions of our controlling anything. Our second nature, whereby we encounter what matters to us in the space of reasons, is both an inescapable feature of our species being and a source of inevitable paradox and conflict. This is the human condition and it is not to be avoided: we can't stop being valuers. Nevertheless, living wholly in the space of reasons—as if our second nature was *all* our nature, as if the question "what value?" could make sense in relation to *everything* that matters to us—is travestying our fundamental life status by radically misrepresenting the absolutely indefeasible claims of what constitutively grounds it. If the cynic, in Wilde's apothegm, knows the price of everything and the value of nothing, *we* recognise the value of everything and the absolute claim—what might, in a now superseded language, have been called the sacredness—of nothing.

That is a life mode which is profoundly unsustainable (to reclaim a crucial term from its current misuse), and the breakdown reaction when it comes will explode from somewhere far deeper in human beings than the claims of reason, never mind those of Enlightenment morality. That reaction will be triggered, I suspect, by two coinciding popular recognitions—by its becoming terrifyingly apparent even to (shall we say?) readers of the *Daily Mail*, firstly, that climate-driven disaster is not after all hot air from crazy environmentalists but is really upon us, and secondly that what our progressivist arrogance has been doing all along in bringing it about is *trashing our humanity*: producing not enlightenment in any possible sense of that term, but the neurotic frenzy of megalopolitan living, the churning restless escapism of motion for motion's sake, and all the slick barbarities of the Internet. When threats to human survival (which is what is at stake) from along those two very different axes come into widely felt collision, the forces released will be shattering in ways we can't yet fully imagine.

I honestly don't think we can now stop the tragic career of worldwide destructiveness, degradation and destabilisation which will provoke that explosion. But through it and after it we might learn the attitude which I have elsewhere [6] described as *letting go* of values. This would involve nothing less than emergence from the long human trauma of the Enlightenment. We could perhaps learn to live again in the absolute natural givens which are the non-rational or a-rational (not *irrational*) ground of all practical reasoning, and correspondingly to recognise those of our reasons for action which *are* expressible in value terms as heuristics, the self's exploring for ongoing provisional commitments through which the given is to be negotiated. Of course, for us to be able to do this, very much of our current institutional framework—using that term in its very widest sense—will have to suffer radical change; in fact, what goes by the name of Western civilisation will have to be largely broken by disaster. Looking realistically at our climate and environmental prospects, we can now be pretty sure that it will indeed be broken. We just have to go on in the hope that this will come sooner rather than later, while enough of the biosphere might still be left to us for retrieval of a humanly habitable planet.

But for any robustness in that hope, we must learn to recognise what is now happening in its full tragic dimensions. Understanding how the 'natural capital' concept misleads us could be a start.

References

1. Barbier, E. 2011. *Capitalizing on nature: Ecosystems as natural assets.* Cambridge: Cambridge University Press.
2. Daly, H. 1996. *Beyond growth: The economics of sustainable development.* Boston: Beacon Press.
3. Ekins, P., et al. 2003. A framework for the practical application of the concepts of critical natural capital and strong sustainability. *Ecological Economics* 44 (2–3): 165–185.
4. Foster, J., ed. 1997. *Valuing nature? Economics, ethics and environment.* London: Routledge.
5. Foster, J. 2008. *The sustainability mirage.* London: Earthscan.
6. Foster, J. 2018. On letting go. In *Post-sustainability: Tragedy and transformation*, ed. J. Foster, 168–184. Abingdon: Routledge.
7. Greaves, T., and R. Read. 2015. Where value resides: Making ecological value possible. *Environmental Ethics* 37: 321–340.
8. Holland, A. 1997. Substitutability: Or, why strong sustainability is weak and absurdly strong sustainability is not absurd. *Valuing nature? Economics, ethics and environment*, ed. J. Foster, 119–134. London: Routledge.

9. Hulme, M. 2014. *Can science fix climate change?* Cambridge: Polity.
10. McDowell, J. 1998. Two sorts of naturalism. In *Mind, value and reality*. Cambridge: Harvard University Press.
11. Sellars, W. 1956. Empiricism and the philosophy of mind. In *Minnesota studies in the philosophy of science*, ed. H. Feigl and M. Scriven, vol. 1, 253–329. Minneapolis: University of Minnesota Press.
12. Thompson, M. 1995. The representation of life. In *Virtues and reasons: Philippa foot and moral theory*, ed. R. Hursthouse, G. Lawrence, and W. Quinn, 247–286. Oxford: Clarendon Press.

CHAPTER 13

The Role of "Natural Capital" in the Debate About Biodiversity

Victor Anderson

Abstract Discussion about the concept of "natural capital" raises many important issues. But is it the most important debate we could be having about the current extinction crisis? I argue that the whole debate about "natural capital" and its monetary valuation—both for and also against—leaves the most important questions about biodiversity, extinction, and the deterioration of ecosystems basically untouched. Particularly significantly, it does not address the biggest economic forces at work: the economics of land use change and the economics of climate change.

The debate about the concept of "natural capital" is an important one and raises many important issues, including philosophical problems raised by the idea of "intrinsic value" and political issues about "commodification". But is it the most important debate we could be having? I will argue that the whole debate about "natural capital" and its valuation—both for and also against—leaves the most important questions about biodiversity, extinction, and the deterioration of ecosystems basically untouched. Particularly significantly, it does not address the main economic forces at work.

V. Anderson (✉)
Global Sustainability Institute, Anglia Ruskin University, Cambridge, UK
e-mail: Victor.anderson@anglia.ac.uk

V. Anderson (ed.), *Debating Nature's Value*,
https://doi.org/10.1007/978-3-319-99244-0_13

The idea of "natural capital" did not originate from natural science or the study of nature. It originated from environmental economics, as an attempt to consider the problem of the deterioration of the natural world through the lens and conceptual structure of marked-based economics. This is a limited aim it achieves fairly thoroughly, using the pairing of "stock" (natural capital) and "flow" (ecosystem services) and applying a variety of economic techniques for monetary valuation.

Three main sorts of objections have been raised and are discussed elsewhere in this book. One is about commodification and entrusting the natural world to the market and market-based ways of thinking. A second objection is based on the idea of "intrinsic value". A third area of objection is that monetary valuation is in practice very difficult to do (for a whole list of reasons); the figures arrived at are therefore always contested, and as a result "valuing natural capital" turns out to be simply a metaphor or a sophisticated sort of slogan rather than a practical methodology.

These are all strong reasons for questioning the validity of "natural capital" as a concept, but I want to approach these issues from a different angle: what if? What if the idea of "natural capital" is accepted and valuation figures can be established and agreed on? What happens then?

I will take as an example, and it is an important one, the destruction of rainforests in Indonesia as a result of mining and plantations. This is destroying biodiverse wildlife habitat and adding to climate change. Indonesia has one of the world's three largest areas of tropical forest, along with Amazonia and Congo. Suppose someone did the calculations and handed the results to the Indonesian Government (and this has been tried), proving that the forests have a greater value left standing than can be extracted by chopping them down and selling the timber, coal, palm oil, and so on that are currently being exported from Indonesia.

I doubt that this would persuade them to change course. They would see the economists' figures as interesting and those who provided the figures probably as well-meaning, but then ask: where is this money? If we could really get what is currently this theoretical money, our plans and policies would change. But we have people to feed and an economy to develop, and we know that selling timber, coal, and palm oil actually brings money in. It is not purely theoretical.

In other words, valuation by itself does not get the job done. The next stage would be for environmental campaigners, and maybe economists, to then try to persuade people—as consumers, taxpayers, voters, investors—to come up with the money and be prepared to pay for the ecosystem services which we in the West currently receive from Indonesia for free.

Two different figures would be important here, more important than the theoretical valuation figures. One would be how much the receivers of the forest ecosystem services, predominantly in the West, would actually be prepared to pay. The other is how much the Indonesians would be willing to accept. In practice, if it ever got this far, there would be a persuasion and bargaining process, rather than the simple application of economists' "natural capital" figures.

We are therefore back to economics and politics. What matters here is the economics of money actually changing hands, as distinct from being a set of figures in a report, and the politics of what people, governments, and companies in both "developed" and "developing" countries can be persuaded to support, pay for, and accept.

"Natural capital" valuation gets us out of none of the problems. It might be used as evidence in the political process, and some people might be persuaded by it, although other forms of evidence and appeal (e.g. based on intrinsic value or emotional attachment) might persuade more people more effectively. However, any such persuasion is in the context of politics, ideas, and economic forces.

I would like us to turn more of our attention to the real economic forces involved here, which are far more powerful than any theoretical monetary valuation calculations. I would also like us to turn our attention to the actual causes of biodiversity loss, mass extinction, and the deterioration of ecosystems. Only by doing that can we put discussion about the concept of "natural capital" into perspective.

Fortunately there is a vast literature on these questions about causes. As far as the immediate causes and outlook are concerned, there appears to be agreement amongst scientists[1] regarding the following:

[1] See, for example:

O.E. Sala et al.: 'Global biodiversity scenarios for the Year 2100', in 'Science' April 2000. http://www.researchgate.net/profile/Rik_Leemans/publication/12606072_Global_biodiversity_scenarios_for_the_year_2100/links/09e4151103847508 2b000000.pdf

Millennium Ecosystem Assessment: 'Ecosystems and Human Well-being: Synthesis' (Island Press 2005). Page 16.

'Global Environment Outlook 5' (United Nations 2012). Page 159.

Will Steffen et al.: 'Planetary Boundaries: Guiding development on a changing planet', in 'Science' 15.1.15. Also Supplementary Materials for this article: https://www.sciencemag.org/content/suppl/2015/01/14/science.1259855.DC1/Steffen-SM.pdf. Page 23.

- Land use change directly produces habitat change and is a major factor in changing ecosystems around the world.
- Climate change is already globally impacting biodiversity and ecosystems and is on course to do so far more in the future.
- There are also problems in particular parts of the world which derive from overabundance of nitrogen, ocean acidification, water stress, and over-exploitation.

If we then ask what the underlying causes are, there is a "first approximation" readily available. The factors most commonly identified are population growth and gross domestic product (GDP) growth. However, this explanation has only a partial validity. The drawback to identifying these as the underlying factors is essentially the same for each: in reality they combine under their two apparently simple headings a vast variety of different pressures on the biosphere.

Although we can say "population" is a factor, it is obvious that the world's population includes people who put pressures on the biosphere in many different ways and to very different extents. Measurements of per capita carbon emissions, for example, show a very widely unequal distribution across the world. So do measurements of consumption of different natural resources. However, at the same time it is also the case that the larger the total population, the larger the total environmental impact is likely to be. Both are true. Although the population debate has become strangely polarised, it seems clear that total population is a factor, whilst at the same time there are enormous differences in the impacts caused by different sections within that total world population.

Something similar can be said about GDP growth. There is a correlation between GDP and environmental impact, but at the same time GDP includes the production of many different types of goods and services, each with different types and extents of impact. It is therefore necessary to be more specific by looking at two big forces driving biodiversity loss: land use change and climate change.

Although we are all used to seeing the world in terms of a geopolitical competition for territory and influence between major powers and alliances, such as the USA, Russia, and China, there is another competition for territory also taking place which is potentially far more important: the competition for the use of land, dividing up the finite amount of land which exists in the world. The competition is principally between three major uses, plus some other less important ones. The three are "wild land", especially forest; agriculture; and urban uses, such as industry, housing, and transport infrastructure.

The ecosystems which underpin all economic activity depend largely on "wild land"—very much a poor relation in the competition with agriculture and urbanisation, both of which have massive economic forces in their favour. Agriculture has the economic power of the growing demand for food, growing not only because of rising population, but also because more and more people in the world are wanting to eat meat, and meat generally requires more land than crops do. Urbanisation has the economic power of manufacturing and the fast drift of economic migrants out of the countryside into towns and cities. Wild land has no such strong purchasing power to defend its position.

This is about **actual purchasing power** and real economic change, not about theoretical valuations. It follows that what would be required to increase the protection of "wild land" is very large transfers of money from rich temperate countries to poorer tropical countries—not as a form of aid, but as a payment for the services which tropical ecosystems provide, in order to outbid other possible uses for the land. Another option is combined uses for the same land, for example, really green cities or careful farming in the rainforests, but again money would need to change hands for those in order to outbid other uses.

Although climate change and biodiversity are often seen by journalists and politicians as separate subjects, in fact climate change is a major and increasing factor causing biodiversity loss. Fossil fuel emissions are destroying biodiversity and ecosystems, not simply destabilising the global climate.[2] There is also another key connection. Earth processes which absorb carbon—such as within forests—often provide other ecosystem services and are homes for biodiversity. Hence some measures to combat climate change, through boosting the capacity to absorb carbon, can also benefit biodiversity.

All this implies that what is known about the economic drivers of climate change can be imported into the debate about the drivers of biodiversity loss. The economics of energy supply is of course crucial here. Fossil fuel subsidies, such as those maintained by the UK and many other countries, are biodiversity destruction subsidies. Dependence on oil, and its rapid

[2] See Richard Pearson: 'Driven to Extinction: the impact of climate change on biodiversity' (Natural History Museum 2011).

rate of consumption, is undermining the competing form of dependence which exists at the same time on a variety of services provided by ecosystems. Failure to invest in the transition to renewable energy is therefore also in practice a failure to invest in ecosystems.

Taking all this into account, there is a clear contrast to be drawn here between two different forms of economics. One is the analysis of the economic forces which drive land use change and climate change. These include the production and consumption of food and energy. The other, very different, but also economics, is the theoretical analysis of how much "natural capital" is worth in terms of money.

Whilst "natural capital" valuation may have some power to persuade, policy debates about food, energy, transport, population, economic growth, urbanisation, and so on are far more important for the future of biodiversity and ecosystems.

If this is so, why is so much of the attention in discussion of the economics of biodiversity given to "natural capital" valuation? How has it come to take the place of discussion about the real economic forces driving biodiversity loss? If you doubt that this has happened, try googling "biodiversity economics" or "economics of biodiversity".

The reasons for this constitute significant barriers to understanding the causes of biodiversity loss. These reasons therefore stand in the way of seriously addressing the mass extinction crisis we find ourselves in. A whole set of reasons, from different parts of the political spectrum, are locked together in a formidable combination.

(1) The real economic drivers of biodiversity loss are highly politically controversial. They include, for example, the role of the oil companies, the workings of the food industry, mining, corporate governance, and international trade arrangements. Many funders of research and conservation do not like to deal with these issues. "Natural capital" valuation feels politically so much safer. This applies generally to the attitudes of trusts, foundations, companies, and most individual donors, and also to the legal limitations created by charity laws, and often also the attitudes of academic research funding bodies.
(2) Within academia, there is a great deal of talk about "interdisciplinarity". However, in practice, education, career, and institutional structures remain organised on a disciplinary basis. Even where interdisciplinarity is achieved within the natural sciences or the

social sciences, it is rarely sufficiently wide to encompass both those fields, and still less often to also include the humanities. Yet the economics of biodiversity loss is inherently interdisciplinary: it requires natural science analysis of what is happening, economic analysis of its causes, historical accounts of how things have been unfolding in the long run, and some philosophical sensitivity to the conceptual issues involved.

(3) As is clear from climate change science and its denial, depressing news is simply unwelcome, particularly where it challenges comfortable views of the world. It is comforting if we can believe that biodiversity loss is just about pandas, tigers, and rhinos rather than about the decline of the natural world as a whole.

(4) Perhaps less easy to explain is the low priority traditionally given to biodiversity issues on the Left. It might seem that the Left would be the place politically most likely to give a warm welcome to critical accounts of what current economic forces and arrangements are doing to the world. Although there are some exceptions (e.g. amongst "ecosocialist" writers), in general this has not been the case for biodiversity and ecosystems. There remains suspicion because of the association of nature with Romanticism and "irrationalism", problems about the association of the defence of the countryside with the aristocracy and conservatism, and difficulty about anything that is "cuddly", "furry", or in some other respect a distraction from "bread and butter issues" or the hard realities of political struggle. The Left has taken climate change on board as an issue, and parts of the Left have concerned themselves with the impact of mining or water privatisation, for example, on communities in developing countries. But it is unusual to find amongst the Left in Western countries a thorough engagement with how the current economy is destroying important parts of the global web of life.

(5) The final problem to mention here is the sheer complexity of the issues involved. The real economics of ecosystems and biodiversity are not simple. This is a difficult analysis to construct in detail, and it can also be difficult to communicate it. Calls for saving the tiger look simple by comparison, and saying that there is a need to value "natural capital" is on the surface a comparatively simple response as well (although in fact very difficult to do in practice).

What follows from all this is that there is not only a set of factors making for biodiversity loss, but also a further set of factors constituting barriers to understanding, or wanting to understand, what is going on. The debate about the concept of "natural capital" is not a great deal of help in efforts to come to terms with either set of factors. Yet it looms large in the economics of biodiversity debate, undeservedly large in my view. The debate about the concept of "natural capital" has turned into a sort of cuckoo in the nest, sitting in a place where a real debate about the economic forces causing extinction and ecosystem deterioration ought to be.

Index[1]

[1] Note: Page numbers followed by 'n' refer to notes.

V. Anderson (ed.), *Debating Nature's Value*,
https://doi.org/10.1007/978-3-319-99244-0

The manufacturer's authorised representative in the EU is Springer Nature Customer Service Centre GmbH, Europaplatz 3, 69115 Heidelberg, Germany. If you have any concerns regarding our products, please contact ProductSafety@springernature.com

Printed and bound by CPI Group (UK) Ltd, Croydon, CR0 4YY
22/09/2026
02221067-0015